ESSENTIAL SKILLS
FOR GCSE
Chemistry

Nora Henry

HODDER EDUCATION
AN HACHETTE UK COMPANY

The Publishers would like to thank the following for permission to reproduce copyright material.

Acknowledgements

Every effort has been made to trace all copyright holders, but if any have been inadvertently overlooked, the Publishers will be pleased to make the necessary arrangements at the first opportunity.

Although every effort has been made to ensure that website addresses are correct at time of going to press, Hodder Education cannot be held responsible for the content of any website mentioned in this book. It is sometimes possible to find a relocated web page by typing in the address of the home page for a website in the URL window of your browser.

Hachette UK's policy is to use papers that are natural, renewable and recyclable products and made from wood grown in well-managed forests and other controlled sources. The logging and manufacturing processes are expected to conform to the environmental regulations of the country of origin.

Orders: please contact Bookpoint Ltd, 130 Park Drive, Milton Park, Abingdon, Oxon OX14 4SE. Telephone: +44 (0)1235 827827. Fax: +44 (0)1235 400401. Email education@bookpoint.co.uk Lines are open from 9 a.m. to 5 p.m., Monday to Saturday, with a 24-hour message answering service. You can also order through our website: www.hoddereducation.co.uk

ISBN: 978 1 510 46001 0

Cover photo © kotoffei - stock.adobe.com

Illustrations by Integra Software Services Pvt. Ltd., Pondicherry, India

Typeset by Integra Software Services Pvt. Ltd., Pondicherry, India

Printed by Replika Press Pvt. Ltd., Haryana, India

A catalogue record for this title is available from the British Library.

Contents

How to use this book

Welcome to *Essential Skills for GCSE Chemistry*. This book covers the major UK exam boards for Science: AQA, Edexcel (including Edexcel International GCSE), OCR 21st Century and Gateway, WJEC/Eduqas and CCEA. Where exam board requirements differ, these specifics are flagged. This book is designed to help you go beyond the subject-specific knowledge and develop the underlying essential skills needed to do well in GCSE Science. These skills include Maths, Literacy, and Working Scientifically, which now have an increased focus in recent years.

- The Maths chapter covers the five key areas required by the government, with different Chemistry-specific contexts. In your Chemistry exams, questions testing Maths skills make up 20% of the marks available.

- The Literacy chapter will help you learn how to answer extended response questions. You will be expected to answer at least one of these per paper, depending on your specification and they are usually worth six marks.

- The chapter on Working Scientifically covers the four key areas that are required in all GCSE sciences.

- The Revision chapter explains how to improve the efficiency of your revision using retrieval practice techniques.

- Finally, the Exam Tips chapter explains ways of improving your performance in the actual exam.

To help you practise your skills, there is an exam-style paper at the end of the book, with another available online at www.hoddereducation.co.uk/EssentialSkillsChemistry. While they are not designed to be accurate representations of any particular specification or exam paper, they are made up of exam-style questions and will require you to put your maths, literacy and practical skills into action.

Key Features

In addition to Key term and Tip boxes throughout the book, there are several other features designed to help you develop your skills.

A Worked example

These boxes contain questions where the working required to reach the answer has been shown.

A Expert commentary

These sample extended responses are provided with expert commentary, a mark and an explanation of why it was awarded.

B Guided questions

These guide you in the right direction, so you can work towards solving the question yourself.

B Peer assessment

These activities ask you to use a mark scheme to assess the sample answer and justify your score.

C Practice questions

These exam-style questions will test your understanding of the subject.

C Improve the answer

These activities ask you to rewrite a sample answer to improve it and gain full marks.

Answers to all questions can be found at the back of the book. These are fully worked solutions with step-by-step calculations included. Answers for the second online exam-style paper can also be found online at www.hoddereducation.co.uk/EssentialSkillsChemistry.

★ **Flags like this one will inform you of any specific exam board requirements.**

1 Maths

Maths skills are included in the assessment of all GCSE science qualifications. At least 20% of the marks awarded in GCSE Chemistry papers will require the use of maths skills. These skills will be tested in the context of Chemistry, though the underlying skills will be same ones you use in maths. For example, you may be asked to calculate a percentage, record data to a number of significant figures or plot a graph. This section aims to take you through all the maths skills needed in order to successfully complete your GCSE Chemistry course.

›› Units and abbreviations

Laboratory work is central to the study of Chemistry, and throughout your course there will be many opportunities for you to use practical work to investigate, record and process data. Some measurements recorded in experimental work may be qualitative and would not include a numerical value. Observations such as 'bubbles' or 'the colourless limewater changed to cloudy' are qualitative measurements, as they refer to the appearance rather than the quantity of things.

Other measurements recorded in experimental work are quantitative. Some of the quantitative measurements and the units you may need to use are shown in Table 1.1. You should familiarise yourself with both the units and their abbreviations.

> **Key terms**
>
> Qualitative: Non-numerical descriptions of how something appears or reacts.
>
> Quantitative: Measurements, such as mass, temperature and volume, involve a numerical value. For these quantitative measurements, it is essential that units are included because stating that the mass of a solid is 0.4 says very little about the actual mass of the solid — it could be 0.4 g or 0.4 kg.

Table 1.1 Units used in Chemistry

Quantitative measurement	Units
Temperature	degrees Celsius (°C)
Area	millimetres squared (mm^2), centimetres squared (cm^2), metres squared (m^2)
Volume	centimetres cubed (cm^3), decimetres cubed (dm^3), metres cubed (m^3)
Mass	milligrams (mg), grams (g), kilograms (kg), tonnes (t)
Time	minutes (min), seconds (s)

Compound units are those composed of more than one unit. For example:

- concentration is measured in mol/dm^3 (mol per dm^3)
- energy change is measured in kJ/mol (kilojoule per mol)

Converting between units

Addition and subtraction of values can only be done if they are expressed in the same units. For example, the mass of an evaporating basin (24 g) cannot be added to the mass of copper oxide (3000 mg) to give a total mass, as the units are different. If the units are different they must be converted to a common unit before being added together. In this example, the mass of copper oxide must first be converted from 3000 mg to 3 g and then added to the mass of the evaporating basin in grams (24 g) to give a total mass of 27 g.

You will need to be able to convert between different volume and mass units as outlined next.

Volume

Volume is usually measured in centimetres cubed (cm^3), decimetres cubed (dm^3) or metres cubed (m^3).

$$1000\,cm^3 = 1\,dm^3$$

$$1000\,dm^3 = 1\,m^3$$

You need to be able to convert between volume units, particularly for calculations on solution volume and concentration. The flow scheme in Figure 1.1 will help you to convert between volume units.

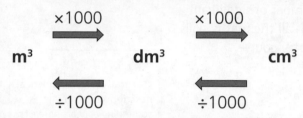

▲ Figure 1.1 Converting between volume units

Tip

For more information on calculating volume, see pages 46–47.

Mass

Mass can be measured in milligrams (mg), grams (g), kilograms (kg) and tonnes (t).

$$1\,tonne = 1000\,kg$$

$$1\,kilogram = 1000\,g$$

$$1\,gram = 1000\,mg$$

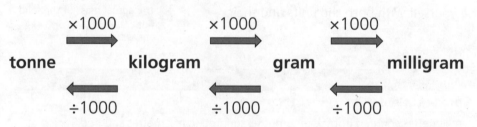

▲ Figure 1.2 Converting between mass units

Tip

Use common sense when converting between units. For example, as a kilogram is bigger than a gram, you should expect to get a larger number of grams when converting from kilograms to grams.

(A) Worked examples

1 a **Convert 35 cm^3 to dm^3.**

Look at Figure 1.1. To convert from cm^3 to dm^3, divide by 1000

$$\frac{35}{1000} = 0.035\,dm^3$$

b **Convert 1.5 dm^3 to cm^3.**

Look at Figure 1.1. To convert from dm^3 to cm^3, multiply by 1000

$$1.5 \times 1000 = 1500\,cm^3$$

c **Convert 325 mg to grams.**

Look at Figure 1.2. To convert from mg to g, divide by 1000

$$\frac{325}{1000} = 0.325\,g$$

d **Convert 4.3 kg to grams.**

Look at Figure 1.2. To convert from kg to g, multiply by 1000

$4.3 \times 1000 = 4300$ g

e **Convert 2.2 tonnes to grams.**

Two conversions are needed.

tonne $\rightarrow$ kilogram $\rightarrow$ gram

Step 1 To convert from tonnes to kilograms, multiply by 1000

$2.2 \times 1000 = 2200$ kg

Step 2 To convert from kilograms to grams, multiply by 1000

$2200 \times 1000 = 2\,200\,000$ g

> **Tip**
> Refer to Figure 1.2 to help with this conversion.

B Guided questions

1 **Convert 1.2 dm³ to cm³.**

To convert from dm³ to cm³, multiply by 1000

$1.2 \times 1000 = $ ....cm³

2 **Convert 8.2 tonnes to g.**

Two conversions are needed here.

tonnes $\rightarrow$ kg $\rightarrow$ g

Step 1 To convert from tonnes to kg, multiply by 1000

$8.2 \times 1000 = $ ....kg

Step 2 Then, to convert from kg to g, multiply by 1000

.... $\times 1000 = $ ....g

> **Tip**
> Always sense-check your answer; a dm³ is a larger unit than a cm³ so you should expect to get a larger number of cm³ when converting.

C Practice questions

3 Carry out the following unit conversions.

a 1.2 dm³ to cm³   **d** 4.4 t to g
b 420 cm³ to dm³   **e** 4 kg to g
c 3452 cm³ to dm³   **f** 3512 g to kg

Making calculations involving conversion of units

There are several topic areas in which you may need to convert units. For example:

- the reaction volumes of gases, where you may need to calculate moles of gas using the equation:

$$\text{amount (in moles)} = \frac{\text{volume (dm}^3)}{24}$$

- the amounts of substance, where you may need to calculate moles using the equations:

$$\text{amount (in moles)} = \frac{\text{mass (g)}}{A_r} \quad \text{or} \quad \text{amount (in moles)} = \frac{\text{mass (g)}}{M_r}$$

> **Tip**
> A_r is the relative atomic mass and is found on the Periodic Table for each element. M_r is the relative formula mass. These values do not have units.

> **Key term**
> Relative formula mass, M_r: The sum of the relative atomic masses (A_r) of all the atoms shown in the formula.

A Worked example

1 Calculate the amount, in moles, present in 2.4 tonnes of magnesium.

To calculate the amount in moles, use the following equation:

$$\text{amount (in moles)} = \frac{\text{mass (g)}}{A_r}$$

Before using this expression, the mass of magnesium must be converted from tonnes to grams.

tonnes → kilograms → grams

$\times 1000 \quad \times 1000$

Step 1

Mass of magnesium in grams $= 2.4 \times 1000 \times 1000 = 2\,400\,000\,g$

Step 2 Amount (in moles) $= \dfrac{\text{mass (g)}}{A_r}$

$$= \frac{2\,400\,000}{24} = 10\,000\,\text{mol}$$

Tip

Some examination boards may ask you to calculate the number of moles rather than the amount in moles – this is answered in the same way.

Tip

Use the periodic table to find the A_r of magnesium. The A_r is 24.

B Guided questions

1 Calculate the amount in moles present in 9.8 kg of sulfuric acid H_2SO_4, which has relative formula mass (M_r) 98.

To calculate the amount in moles use the expression:

$$\text{amount (in moles)} = \frac{\text{mass (g)}}{M_r}$$

Step 1 Convert the mass from kg to g by multiplying by 1000

9.8 × 1000 = ...

Step 2 Substitute the mass in grams and the M_r into the equation to calculate your final answer.

$$\text{amount (in moles)} = \frac{\text{mass (g)}}{98} = \ldots\ldots\ldots = \ldots\ldots\ldots$$

2 Calculate the amount in moles present in 48 000 cm³ of nitrogen gas.

To calculate the amount in moles of a gas use the expression:

$$\text{amount (in moles)} = \frac{\text{volume (dm}^3)}{24}$$

Step 1 Convert the gas volume from cm³ to dm³ by dividing by 1000

$$\frac{48\,000}{1000} = \ldots\ldots\ldots \text{dm}^3$$

Step 2 Substitute the volume into the equation and calculate your final answer.

$$\text{amount (in moles)} = \frac{\text{volume (dm}^3)}{24} = \ldots\ldots\ldots = \ldots\ldots\ldots$$

Tip

This equation can only be used to calculate the moles of a gas if the volume is given.

C Practice questions

3 Calculate the amount in moles present in 6 kg of calcium (Ca).

4 Calculate the amount in moles present in 3.2 tonnes of calcium carbonate. Calcium carbonate has relative formula mass (M_r) = 100.

5 Calculate the amount in moles present in 17 kg of ammonia (NH_3).

6 Calculate the amount in moles present in 2.1 tonnes of iron(III) oxide (Fe_2O_3).

7 Calculate the amount in moles present in 0.592 kg of magnesium nitrate ($Mg(NO_3)_2$).

8 Calculate the amount in moles present in 7200 cm³ of sulfur trioxide gas (SO_3).

» Arithmetic and numerical computation

Expressions in decimal form

When adding or subtracting data, decimal places are often used to indicate the precision of the answer. The term 'decimal place' refers to the numbers after the decimal point. The number of decimal places is the number of digits after the decimal place.

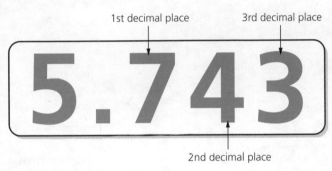

▲ Figure 1.3 Decimal places

The number 5.743 has 3 decimal places, while the number 10 has no decimal places.

Sometimes in calculations you are asked to present your answer in a calculation to 1 or 2 decimal places. To do this, you need to round your answer. For example:

● rounding a number to 1 decimal place means there is only one digit after the decimal place.
● rounding a number to 2 decimal places means there are two digits after the decimal place.

To round a number to a given number of decimal places, look at the digit after the last one you need and

● if the next number is *5 or more*, round up
● if the next number is *4 or less*, do not round up.

For example, if you are rounding a number to 2 decimal places, it is useful to underline all numbers up to two numbers after the decimal place. This then focuses your attention on the next number, which helps with rounding.

A Worked examples

1 a Round 2.2625 g to 2 decimal places.

Step 1 Underline all the numbers up to two numbers after the decimal point.

2.26<u>25</u>

Step 2 Look at the number after the last underlined number. This number is 2 so, the rule 'if the next number is 4 or less, do not round up' applies here and the number 6 is unchanged.

The answer is 2.26 g (to 2 d.p.)

b Round 4.9762 g to 1 decimal place.

Step 1 Underline all the numbers up to one number after the decimal point.

4.9<u>7629</u>

Step 2 Look at the number after the last underlined number. This number is 7 so, the rule 'if the next number is 5 or more, round up' is followed. This means the number 9 is rounded up to 10.

The answer is 5.0 g (to 1 d.p.)

B Guided questions

1 Round 3.418 g to 2 decimal places.

Step 1 Underline all the numbers up to two numbers after the decimal point.

3.41<u>8</u>

Step 2 Now look at the number after the last underlined number and decide, by using rounding rules, if you need to round up or not.

The answer is .... (to 2 d.p.)

2 In an experiment to find the mass of water removed on heating a solid, a student recorded the following measurements:

Mass of hydrated **solid + evaporating basin = 28.465 g**

Mass of evaporating basin = 26.250 g

Mass of anhydrous **solid + evaporating basin = 27.799 g**

a Calculate the mass of the anhydrous solid to 2 decimal places.

Step 1 Subtract the mass of the evaporating basin from the mass of the evaporating basin + anhydrous solid.

27.799 − 26.250 = 1.549 g

Step 2 Underline the numbers up to 2 numbers after the decimal place.

....

Step 3 Look at the number after the last underlined number and decide, using rounding rules, whether you need to round up or not.

....

> **Key terms**
>
> Hydrated: A substance that contains water.
>
> Anhydrous: A substance that does not contain water.

b **Calculate the mass of water removed to 2 decimal places.**

Step 1 Subtract the combined mass of the anhydrous solid + evaporating basin from the combined mass of the hydrated solid + evaporating basin.

$28.465 - 27.799 = $ ....g

Step 2 Underline the numbers up to 2 numbers after the decimal place.

....

Step 3 Look at the number after the last underlined number and decide, using rounding rules, whether you need to round up or not.

....

C Practice questions

3 In an experiment, different solids were heated in evaporating basins. To determine the mass of solid after heating, the mass of the evaporating basin must be subtracted from the mass of solid and evaporating basin. Calculate the mass of solid to 1 decimal place by carrying out the following subtractions.

a $30.25\,g - 28.53\,g$ c $24.34\,g - 22.23\,g$

b $35.67\,g - 25.98\,g$

4 Copper(II) sulfate solution was electrolysed for 5 minutes using copper electrodes. The table shows the mass of the copper anode and cathode before and after the electrolysis.

	anode	cathode
Mass of electrode before electrolysis/g	1.66	1.58
Mass of electrode after electrolysis/g	1.15	1.87

Calculate the mass of copper deposited to 1 decimal place.

5 Copy and complete the table.

Mass/g	Mass recorded to 2 decimal places/g
29.883	
0.046	
32.6789	
13.999	
0.0894	
19992.456	

Tip
In this electrolysis experiment, copper is deposited on one electrode that gets heavier and the other electrode gets lighter.

Recording to an appropriate number of decimal places

When adding or subtracting measurements with different numbers of decimal places, the accuracy of the final answer should be no greater than the least accurate measurement. This means that, when measurements are added or subtracted, the answer should have the same number of decimal places as the smallest number of decimal places in any number involved in the calculation.

A Worked example

1 A student recorded the following masses of potassium chloride: 32.23 g, 2.1 g and 4.456 g. Calculate the total mass of potassium chloride. Give your answer to the appropriate number of decimal places.

Step 1 Add the masses.

32.23 + 2.1 + 4.456 = 38.786 g

Step 2 Look at each of the masses and record the number of decimal places in each.

Measurement	Number of decimal places
32.23 g	2
2.1 g	1
4.456 g	3

The number with the least decimal places is 2.1 – 2.1 has only one decimal place, so the answer must be rounded to one decimal place.

Step 3 Underline the numbers up to 1 number after the decimal place.

38.786

Step 4 Work out which way to round – up or down. The number after the last underlined number is 8. This is greater than 5 so the answer must be rounded up.

The answer is 38.8 g.

> **Tip**
> In a calculation using different measurements, the answer is recorded to the same accuracy as the measurement with the smallest number of decimal places.

B Guided question

1 A student recorded the initial temperatures of different solutions in an experiment. The results are shown in the table. Calculate the mean temperature. Give your answer to the appropriate number of decimal places.

Step 1 Record the number of decimal places in each reading on a copy of the table here.

Temperature/°C	Number of decimal places
10.25	
10	
10.2	
10.0	

Step 2 Identify the number of decimal places in the temperature with the least accuracy. This is the number of decimal places that should be in your answer. The answer should be recorded to 0 decimal places.

....

Step 3 Calculate the mean temperature and record to the correct number of decimal places.

$$= \frac{\ldots\ldots\ldots\ldots\ldots\ldots\ldots}{4} = \ldots\ldots\ldots\ldots\ldots\ldots\ldots$$

> **Tip**
> To calculate a mean value, add all the values and divide by the total number of values.

C Practice questions

2 A technician weighed out three different masses of each chemical and recorded the following masses. Work out the total mass of each chemical, giving your answer to the appropriate accuracy.

    a Calcium carbonate 43.2 g, 0.245 g, 10.222 g
    b Copper(II) sulfate 0.245 g, 10.393 g, 2.49 g
    c Copper(II) oxide 3.23 g, 0.3439 g, 3.97 g

3 Copy and complete the table.

Mass of evaporating basin and solid/g	Mass of evaporating basin/g	Mass of solid (to appropriate number of decimal places)/g
34.567	23.4	
29.93	25.66	
25.49	22.1	

Expressions in standard form

Standard form is used to express very large or very small numbers so that they are more easily understood and managed. It's easier to say that a speck of dust weighs 1.2×10^{-6} grams than to say it weighs 0.000 001 2 grams, or that a carbon-to-carbon bond has a length 1.3×10^{-10} m than to say it is 0.000 000 000 13 m.

Standard form must always look like this:

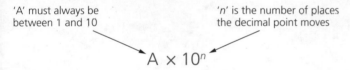

'A' must always be between 1 and 10

'n' is the number of places the decimal point moves

$$A \times 10^{n}$$

▲ **Figure 1.4** Standard form

'n' can also be thought of as the power of ten that A is multiplied by to equal the original number. It is important that you can convert standard form back to ordinary form and from ordinary form back to standard form.

The International System of Units

The International System of Units (SI) is a system of units of measurements that is widely used all over the world. It uses several base units of measure; for example, metres, grams and seconds. In the SI system there is a set of prefixes and when these are used, the power of ten standard notation is avoided. A prefix always goes before the unit itself. Table 1.2 gives the commonly used prefixes and their meanings.

Table 1.2 Commonly used prefixes in SI

Prefix name	Symbol	Standard form	decimal
tetra-	T	$\times 10^{12}$	1 000 000 000 000
giga	G	$\times 10^{9}$	1 000 000 000
mega	M	$\times 10^{6}$	1 000 000
kilo	k	$\times 10^{3}$	1000
centi	c	$\times 10^{-2}$	0.01
milli	m	$\times 10^{-3}$	0.001
micro	μ	$\times 10^{-6}$	0.000 000 1
nano	n	$\times 10^{-9}$	0.000 000 001

A Worked examples

1 **Write 3 450 000 in standard form.**

   Step 1 Write the non-zero digits with a decimal place after the first number and then write $\times 10^n$ after it.

   3.45×10^n

   Step 2 Count how many places the decimal point has moved and write this value as the n value. The n value is positive as 345 000 is greater than 1.

   $3 450 000 = 3.45 \times 10^6$

2 **Write 0.000 947 in standard form.**

   Step 1 Write the non-zero digits with a decimal place after the first number and then write $\times 10$ after it.

   9.47×10^n

   Step 2 Count how many places the decimal point has moved and write this value as the n value – the n is negative because 0.000 947 is less than 1.

   $0.000947 = 9.47 \times 10^{-4}$

3 **Atoms have a radius of 0.1 nm. Write this in standard form in metres.**

   Step 1 Write down the conversion $1 nm = 1 \times 10^{-9}$

   Step 2 Work out how many nanometres 0.1 is by using this equation:

   $0.1 nm = 0.1 \times 1 \times 10^{-9} = 1 \times 10^{-10} m$

4 **A coarse particle has a diameter of 0.000 002 5 m. Write this diameter using a prefix.**

   Step 1 Write the number in standard form.

   $2.5 \times 10^{-6} m$

   Step 2 Look at Table 1.6. Notice that 10^{-6} has the prefix micro.

   $2.5 \times 10^{-6} m = 2.5 \ \mu m$

> **Tip**
> Negative powers mean the number is less than one. Positive powers mean the number is greater than 1.

> **Tip**
> Make sure you can enter standard form on your calculator.

B Guided questions

1 **Write 0.006 45 in standard form.**

   Step 1 Write the non-zero digits with a decimal place after the first number and then write '$\times 10^n$' after it.

   6.45×10^n

   Step 2 Count how many places the decimal point has moved and write this value as the n value. It is a negative n value as 0.006 45 is less than 1.

   ....

2 **Calculate the number of atoms present in 2.3 g of sodium.**

   Step 1 Calculate the amount in moles of sodium using:

   $$\text{Amount (in moles)} = \frac{\text{mass (g)}}{M_r} = \frac{2.3}{23} = 0.1$$

   Step 2 One mole of atoms contains 6.02×10^{23} atoms. To find the number of atoms present in 2.3 g of sodium multiply the amount in moles by 6×10^{23}

   ....

> **Tip**
> A number that you will use in Chemistry, which is usually presented in standard form, is Avogadro's number, it is 6.02×10^{23}.

> **Key term**
> Avogadro's number: The number of atoms, molecules or ions in one mole of a given substance.

C Practice questions

3 Write the following numbers in standard form.
   a 11 345
   b 0.01
   c 0.003 45

4 Write the following numbers in ordinary form.
   a 3.2×10^6
   b 8.456×10^2
   c 5.6765×10^{-3}

5 Use your calculator to calculate the following. Write your answer in standard form to 2 decimal places.
   a $6.02 \times 10^{23} \times 2.20 \times 10^{-12}$
   b $0.000034 \times 0.3333343$
   c $6.02 \times 10^{23} \div 2.10 \times 10^8$

6 Carbon atoms have a radius of 0.070 nm. Write this in standard form in metres.

7 An atom of hydrogen contains a proton and a neutron. Calculate the mass of a hydrogen atom if a proton has mass 1.6725×10^{-24} and an electron has mass 0.0009×10^{-24}.

8 Calculate the number of tin atoms in 238 g of tin.

> **Tip**
> In Questions 7 and 8 you need to use Avogadro's number.

Using fractions and percentages

A fraction is a part of a whole. It can be written as a whole number divided by another whole number.

For example, one-half is a fraction and is written as $\frac{1}{2}$. To calculate the value of a fraction as a decimal, divide the top line by the bottom

$$\frac{1}{2} = 0.5$$

Per cent means 'out of 100'. If 50% of the population own a car, this means 50 out of every 100 people have one. The symbol % means per cent.

Percentages are a useful way of showing proportions. For example, the major uses of ammonia can be expressed as percentages as shown in Figure 1.5.

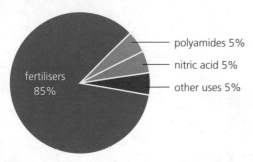

polyamides 5%
nitric acid 5%
other uses 5%
fertilisers 85%

▲ **Figure 1.5** The major uses of ammonia

> **Tip**
> This is an example of a pie chart. For examples of other types of charts see pages 25–27.

A percentage is a fraction out of 100. For example, 36% means 36 out of a hundred and is written as a fraction as $\frac{36}{100}$

36% as a decimal is 0.36

To convert a fraction or decimal to a percentage, simply multiply by 100. For instance, to convert 0.25 to a percentage,

$$0.25 \times 100 = 25\%$$

To convert $\frac{1}{5}$ to a percentage,

$$\frac{1}{5} \times 100 = 20\%$$

To calculate a *percentage change* (this may be an increase or a decrease) use the following equation.

$$\text{Percentage change} = \frac{\text{change in value}}{\text{original value}} \times 100$$

Other types of calculation that involve percentages include percentage yield and atom economy calculations. To calculate these, you need to recall and use the equations shown here.

$$\text{Percentage yield} = \frac{\textit{actual} \text{ yield}}{\text{theoretical yield}} \times 100$$

$$\% \text{ atom economy} = \frac{\text{molecular mass of desired product}}{\text{sum of molecular masses of all reactants}}$$

A Worked examples

1 a The chromatogram produced from a sample of ink is shown below. Calculate the R_f value for the green dye. Give your answer to one decimal place.

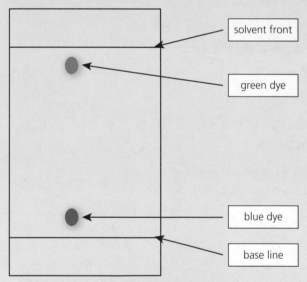

solvent front

green dye

blue dye

base line

Key term

R_f: R_f value in chromatography is the distance travelled by a given component divided by the distance travelled by the solvent.

Step 1 Measure the distance moved by the green dye from the base line. Measure to the middle of the spot.

Distance = 43 mm

Step 2 Measure the distance moved by the solvent from the base line to the solvent front.

Distance = 49 mm

Step 3 Substitute the values into the equation.

$$R_f = \frac{\text{distance moved by spot}}{\text{distance moved by solvent}} = \frac{43}{49} = 0.9$$

Tip

R_f values are less than one.

Tip

You can measure the distances in mm or cm but make sure you record both to the same unit. R_f does not have any units.

b **The amount of carbon dioxide produced from industry has increased from 500 000 tonnes to 750 000 tonnes in a six-year period. The amount produced from electricity generation has increased from 750 000 tonnes to 900 000 tonnes. Calculate the difference in percentage increase in carbon dioxide produced from industry and from electricity generation.**

Step 1 Calculate the change in mass of carbon dioxide by subtracting the two masses.

Industry: change in mass = 750 000 − 500 000 = 250 000

Electricity: change in mass = 900 000 − 750 000 = 150 000

Step 2 Calculate the percentage change in mass by dividing by the original and multiplying by 100

Industry: $\dfrac{250\,000}{500\,000} \times 100 = 50\%$

Electricity: $\dfrac{150\,000}{750\,000} \times 100 = 20\%$

Step 3 Calculate the difference in percentage increase by subtracting the two values.

50 − 20 = 30%

B Guided questions

1 **Calculate the percentage of nitrogen in $Ca(NO_3)_2$.**

Step 1 You first need to find the relative formula mass (M_r) of $Ca(NO_3)_2$

$M_r = 40 + (14 \times 2) + (16 \times 6) = $ ....

Step 2 There are two nitrogen atoms and so the mass of nitrogen in $Ca(NO_3)_2$ is $14 \times 2 = $ ....

Step 3 Express the quantity as a fraction.

$\dfrac{\text{mass of nitrogen}}{M_r} = $ ....

Step 4 Multiply by 100 to express as a percentage.

2 **Pure gold is 24 carat. A ring of mass 6 g is made from 9 carat gold. Calculate the mass of gold in this ring.**

Step 1 Calculate the proportion of gold in the ring.

$\dfrac{9}{24} = $ ....

Step 2 Multiply by the mass of the ring.

Tip

To find the percentage of a quantity:
1 Write the percentage as a fraction.
2 Multiply by the quantity.

C Practice questions

3 A nail has mass 7.45 g. After a month, the nail was rusted and its mass was found to be 7.97 g. Calculate the percentage increase in mass. Give your answer to 2 decimal places.

4 An alloy contains 45 g of lead and 30 g of tin. Calculate the percentage of tin in the alloy.

5 Calculate the percentage by mass of

   a Hydrogen in $Ca(OH)_2$ **c** Nitrogen in $(NH_4)_2SO_4$
   b Potassium in $K_2Cr_2O_7$

6 Some eggshells were found to contain 35% of calcium carbonate. Calculate the mass of calcium carbonate in 2.3 g of egg shells.

7 A sample of 6.7 g of a rock was found to contain 4.1 g of silicon dioxide. What percentage of the rock was silicon dioxide?

Ratios

A ratio is a way to compare amounts of something. Recipes, for example, are sometimes given as ratios. To make pastry you may need to mix 2 parts flour to 1 part fat. This means the ratio of flour to fat is 2:1. Ratios are written with a colon (:) between the numbers and usually only whole numbers are used.

Ratios are similar to fractions; they can both be simplified by finding common factors.

For example, in the ratio 12:15 the highest common factor is 3 so the ratio simplifies to 4:5.

Ratios are used in Chemistry in many calculations, such as in working out empirical formulas, calculating reacting masses and balancing equations.

Tip

A factor is a number that divides exactly into another number.

A Worked example

1 A compound has the molecular formula $C_2H_4O_2$. What is the empirical formula?

To find the simplest ratio, divide the number of moles by the smallest of the numbers of atoms; in this case 2.

$$C : H : O$$
$$\frac{2}{2} : \frac{4}{2} : \frac{2}{2}$$
$$1 : 2 : 1$$

CH_2O is the empirical formula.

Key term

Empirical formula: The simplest whole number ratio of atoms present.

B Guided question

1 A compound contains 0.050 moles of phosphorus and 0.125 moles of oxygen atoms. What is its molecular formula?

Step 1 Write down the elements present and the moles of each underneath.

   P : O

   .... : 0.125

Step 2 To find the simplest ratio divide by the smaller of the number of moles.

$$\frac{0.050}{0.050} : \frac{0.125}{..........}$$

Sometimes, you may not get a whole number ratio at this stage, and often multiplying by 2 or another number is necessary.

C Practice questions

2 Write the empirical formula of the following compounds.

    a $C_{16}H_{20}N_8O_4$ **c** $C_6H_{12}O_6$

    b $Na_2S_2O_3$ **d** P_4O_{10}

3 What is the empirical formula of a compound that contains

    a 0.6 moles of lead atoms and 0.8 moles of oxygen atoms?

    b 1.093 moles of chlorine atoms and 3.825 moles of oxygen atoms?

4 A compound contains 0.207 g of lead and 0.032 g of oxygen. Calculate the empirical formula of the compound.

Balancing equations

In a balanced chemical equation, the substances are all in ratio to each other and this is shown by the numbers in front of each formula in the balanced symbol equation. For example, 2 moles of magnesium react with 1 mole of oxygen to produce 2 moles of magnesium oxide.

$$2Mg + O_2 \rightarrow 2MgO$$

The ratio is

    2 moles Mg : 1 mole O_2 : 2 moles MgO

Or in the reaction

$$2Al + 6HCl \rightarrow 2AlCl_3 + 3H_2$$

The ratio between aluminium and hydrogen is

    2 moles Al : 3 moles H_2

The ratio between aluminium and hydrochloric acid is

    2 moles Al : 6 moles HCl

This can be simplified to

    1 mole Al : 3 moles HCl

Ratios can be used in calculating the number of moles that react together in a reaction.

A Worked example

1 **a** **In the reaction, $2Pb(NO_3)_2$ (s) $\rightarrow 2PbO$ (s) $+ 4NO_2$ (g) $+ O_2$ (g)**

    how many moles of NO_2 are produced from 0.35 moles of $Pb(NO_3)_2$?

    Step 1 Write down the ratio between the two substances using the equation and simplify.

    $Pb(NO_3)_2$: NO_2

           2 : 4

        1 : 2

    Step 2 Apply this ratio to the 0.35 moles of $Pb(NO_3)_2$

    There is twice as much NO_2 as $Pb(NO_3)_2$ so multiply $Pb(NO_3)_2$ moles by 2

    $0.35 : (2 \times 0.35) = 0.7$ mol

b $4Al + 3O_2 \rightarrow 2Al_2O_3$

   i **How many moles of aluminium are needed to produce 0.76 moles of aluminium oxide?**

     Step 1 Write down the ratio between the two substances using the equation and simplify.

       Al : Al_2O_3

       4 : 2

       2 : 1

     Step 2 Apply this ratio to the 0.76 moles of Al_2O_3

     There are twice as many moles of aluminium so multiply by 2

      $0.76 \times 2 = 1.52 \, mol$

   ii **How many moles aluminium oxide are produced from 0.2 moles of aluminium?**

     Step 1 Write down the ratio between the two substances using the equation and simplify.

       Al : Al_2O_3

       4 : 2

       2 : 1

     Step 2 Apply this ratio to the 0.2 moles of aluminium.

     There is half as many moles of aluminium oxide, so divide by 2 $\frac{0.2}{2} = 0.1 \, mol$

B Guided question

1 **In the reaction $N_2 + 3H_2 \rightarrow 2NH_3$**

  how many moles of nitrogen are needed to react fully with 0.4 moles of hydrogen?

  Step 1 Write down the ratio between the two substances using the equation and simplify.

    N_2 : H_2

     1 : 3

  Step 2 Apply this ratio to the 0.4 moles of hydrogen.

  ....

C Practice questions

2 In the reaction $2Cu(NO_3)_2 \, (s) \rightarrow 2CuO \, (s) + 4NO_2 \, (g) + O_2 \, (g)$

  a How many moles of O_2 are produced from 4 moles of $Cu(NO_3)_2$?
  b How many moles of NO_2 are produced from 0.6 moles of $Cu(NO_3)_2$?

3 In the reaction $CaO + 3C \rightarrow CaC_2 + CO$

  a How many moles of carbon are needed to completely react with 0.33 moles of CaO?
  b How many moles of CO are produced when 3.2 moles of carbon react?

4 In the reaction $3Pb + 2O_2 \rightarrow Pb_3O_4$

  a How many moles of oxygen are needed to completely react with 0.66 moles of Pb?
  b How many moles of Pb_3O_4 are produced when 2.2 moles of oxygen react?
  c How many moles of Pb_3O_4 are produced when 0.33 moles of lead react?

Estimating the results of simple calculations

When carrying out calculations it is often useful to estimate the answer first. Estimates give a rough figure for the final answer. In Maths, the word 'Estimate' generally means to round each number to 1 significant figure before performing the calculation. By rounding numbers in a calculation, a rough idea of what the answer should be can be determined quickly. This is useful as you can then check if your final answer obtained using a calculator is likely to be correct, or if you have made an error by comparing it to your estimate.

★ Not explicitly required for WJEC or Edexcel International GCSE Chemistry students.

Tip

Even if a skill is not explicitly required by your exam board, you will likely cover it in Maths GCSE, so it can't hurt to refresh your memory.

A Worked example

1 **Estimate the value of 4.99×6.18**

Step 1 Round each number to 1 significant figure.

$5 \times 6 = 30$

The estimated value is 30. The calculated value using a calculator is 30.84

B Guided question

1 In a chromatography experiment, a student found that a compound moved a distance of 8.2 cm and the solvent moved a distance of 19.6 cm. Estimate the R_f value and use it to decide whether the compound is P, which has an R_f value of 0.4, or Q, which has an R_f value of 0.2.

Step 1 Write down the equation used to calculate R_f

$$R_f = \frac{\text{distance moved by compound}}{\text{distance moved by solvent}}$$

Step 2 Round each distance to 1 significant figure.

compound distance = 8   solvent distance = ...

Step 3 Estimate the R_f value.

....

C Practice questions

2 What is the approximate radius of an atom?

A 0.1×10 m   C 1×10^{-10} m
B 1×10^{-1} m   D 1×10^{-100} m

3 The table shows the boiling point of some halogens.

	Fluorine	Chlorine	Bromine	Iodine
Boiling point/°C	−188	35		185

Estimate the boiling point of bromine.

4 Estimate the value of 98.97×28.98.

5 Estimate the volume of a cube that has the side length of 9.7 mm.

»» Handling data

Significant figures

In some calculations you may get a long decimal answer on your calculator display. In these instances, it is important to round the answer correctly using significant figures. Significant figures are the digits of a number that are used to express it to a given degree of accuracy.

The first significant figure of a number is the first digit that is not a zero.

The rules for significant figures are:

- Non-zero digits are always significant.
- Zeros at the start of a number (leading zeros) are never significant.
- Zeros between two non-zero digits are significant.
- When a number has no decimal point but ends in several zeros, these zeros may or may not be significant. The number of significant figures should then be stated.

▲ Figure 1.6 A very long decimal answer on a calculator can be rounded to significant figures

The rules are best illustrated with some examples:

- 34.23 has *4* significant figures – always count non-zero digits.
- 6000 has no decimal point and ends in several zeros so it is difficult to say if the zeros are significant. With zeros at the end of a number, the *number* of significant figures should be stated.
- 2000.0 has a decimal place, hence it has *5* significant figures.
- 0.036 has *2* significant figures as the leading zeros are not significant.
- 3.0212 has *5* significant figures.

In some calculations, you should round the answer to a certain number of significant figures.

The rules for rounding are:

- If the next number is *5 or more*, round up.
- If the next number is *4 or less*, do not round up.

A Worked example

1 **Calculate the value of 7.799 g – 6.250 g. Give your answer to 3 significant figures.**

$$7.799 - 6.250 = 1.549\,g$$

The final answer is 1.55 g (to 3 s.f.) because the digit 9 is greater than 5.

B Guided question

1 **Write the number 3 478 906 to 3 significant figures.**

Step 1 Underline the first three figures.

<u>347</u>8906

Step 2 Write down the first two figures.

34

Step 3 Look at the number after the one you have underlined and apply the rounding rules.

....

Step 4 Write zero for all the remaining digits.

....

C Practice questions

2 What is the number of significant figures in each of the following numbers?

   a 13.43
   b 0.002 904
   c 2300.0

   d 1.604
   e 0.094

3 Write the following numbers to the stated number of significant figures.

   a 35 561.22 to 4 significant figures
   b 5.278 to 3 significant figures
   c 423 to 1 significant figure

   d 442.45 to 4 significant figures
   e 0.000 045 193 to 2 significant figures.

4 In an experiment the theoretical yield to prepare some copper(II) sulfate crystals was 2.85 g.

$$CuO + H_2SO_4 \rightarrow CuSO_4 + H_2O$$

Only 2.53 g of copper(II) sulfate was obtained.

Calculate the percentage yield of copper(II) sulfate in this experiment and give your answer to 2 significant figures.

5 In an experiment 95.0 cm^3 of oxygen gas was collected. The volume of one mole of any gas at room temperature and pressure is 24.0 dm^3.

How many moles of oxygen is 95.0 cm^3? Give your answer to 3 significant figures.

> **Tip**
> Remember that moles of gas
> $$= \frac{\text{volume of gas (cm}^3)}{24}$$

Reporting calculations to an appropriate number of significant figures

When combining measurements with different degrees of accuracy and precision, the accuracy of the final answer can be no greater than the least accurate measurement. So, when measurements are multiplied or divided, the answer can contain no more significant figures than the least accurate measurement. You may be asked to 'give your answer to the appropriate precision or appropriate number of significant figures'.

A Worked example

1 Calculate the value of $\dfrac{1.84 \times 2.3}{3.02}$. Give your answer to the appropriate precision.

Step 1 Using a calculator, the value is 1.401 324 5.

Step 2 Write down the number of significant figures in each measurement.

Measurement	Number of significant figures
1.84	3
2.3	2
3.02	3

From the table you can see that the least accurate measurement is 2.3, which has 2 significant figures. Hence, your answer should be rounded to 2 significant figures.

1.401 324 5 rounds to 1.4 (to 2 s.f.)

B Guided question

1 In a titration 20.5 cm^3 of 0.25 mol dm^{-3} sodium hydroxide solution reacts with 1.2 mol dm^{-3} hydrochloric acid.

$$NaOH + HCl \rightarrow NaCl + H_2O$$

Calculate the volume of hydrochloric acid required to neutralise the sodium hydroxide solution. Give your answer to the appropriate number of significant figures.

Step 1 Write down the number of significant figures in each measurement in the table.

Measurement	Number of significant figures
20.5 cm^3	
0.25 mol/dm^3	
1.2 mol/dm^3	

Step 2 Write down how many significant figures there are in the least accurate measurement.

....

Step 3 Calculate the number of moles of sodium hydroxide.

$$moles\ NaOH = \frac{volume \times conc.}{1000} = \frac{20.5 \times 0.25}{1000} = \ldots\ldots\ldots\ldots$$

Step 4 Use the ratio from the balanced equation (1 : 1) to write down the moles of HCl.

....

Step 5 Calculate the volume of HCl.

$$volume = \frac{moles \times 1000}{conc.}$$

....

Step 6 Round your answer to the same number of significant figures as the least accurate measurement.

....

C Practice questions

2 Use density = $\dfrac{mass}{volume}$ to calculate the density of a block of iron that has mass of 40.52 g and volume of 5.1 cm^3. Give your answer to an appropriate number of significant figures.

3 Calculate the number of moles in 2.7 g of calcium. Give your answer to an appropriate number of significant figures.

4 In a titration 26.50 cm^3 of 0.200 mol/dm^3 sodium hydroxide solution reacts with 0.300 mol/dm^3 hydrochloric acid.

$$NaOH + HCl \rightarrow NaCl + H_2O$$

Calculate the volume of hydrochloric acid required to neutralise the sodium hydroxide solution. Give your answer to the appropriate number of significant figures.

Significant figures and standard form

For numbers in standard form, to find the number of significant figures ignore the exponent (n number) and apply the usual rules.

For example, 6.2090×10^{28} has five significant figures and 1.3×10^2 has two significant figures.

The same number of significant figures must be kept when converting between ordinary and standard form

$$0.0050 \, mol/dm^3 = 5 \times 10^{-3} \, mol/dm^3 \text{ (to 2 s.f.)}$$

$$40.06 \, g = 4.006 \times 10^1 \, g \text{ (to 4 s.f.)}$$

$$90.0 \, g = 9.00 \times 10^1 \, g \text{ (to 3 s.f.)}$$

$$0.02070 \, kg = 2.070 \times 10^{-2} \, kg \text{ (to 4 s.f.)}$$

The number 260.99 rounded to

4 significant figures is 261.0

3 significant figures is 261

2 significant figures is 260

1 significant figures is 300

Using standard form makes it easier to identify significant figures.

In the example, 261 has been rounded to the 2 significant figure value of 260. However, if seen in isolation, it would be impossible to know whether the final zero in 260 is significant (and the value to 3 significant figures) or insignificant (and the value to 2 significant figures).

Standard form is unambiguous, however, and the number of significant figures is obvious:

2.6×10^2 is to 2 significant figures

2.60×10^2 is to 3 significant figures

(A) Worked example

1 **Convert 0.002 350 to standard form. Keep the correct number of significant figures.**

Step 1 Write down the number of significant figures in 0.002 350.

There are 4, as the leading zeros are not significant.

Step 2 Write out the 4 significant digits with a decimal place after the first one and write $\times 10^n$ after it

2.350×10^n.

Step 3 Count how many places the decimal place has moved and write this as an n value. It is a negative value, as 0.002 350 is less than zero.

2.350×10^{-3}

(B) Guided question

1 **Convert 3209.0 to standard form. Keep the correct number of significant figures.**

Step 1 Write down the number of significant figures in 3209.0.

There are 5.

Step 2 Write the five significant digits with a decimal place after the first one and write $\times 10^n$ after it.

.... $\times 10^n$

Step 3 Count how many places the decimal place has moved and write this as an n value.

....

> **Tip**
> The last digit in 3209.0 is significant and must be retained when converting to standard form.

(C) Practice questions

2 Complete the table to convert the numbers from ordinary to standard form, retaining the number of significant figures.

Number	Number of significant figures	Number in standard form
0.0060		
50.08		
3000.6 g		
0.040 70		

3 Write the following in standard form, keeping the same number of significant figures.

a 0.050

b 0.120

c 1 230 010

d 14 050

e 30.03

Finding arithmetic means

The arithmetic mean is found by adding together all the values and dividing by the total number of values. It may be referred to as the 'average' or simply as the mean.

> **Key term**
> Arithmetic mean: The sum of a set of values divided by the number of values in the set – it is sometimes called the average or mean.

(A) Worked example

1 **a** The temperature of a solution was measured every 30 seconds for 3 minutes and the results recorded in this table.

Time/s	0	30	60	90	120	150	180
Temperature/°C	21	22	23	24	24	23	22

Calculate the mean temperature of the solution over this time period.

In this case, there are 7 values, so the method is to add up the individual values and divide by 7

$$\text{Mean} = \frac{21 + 22 + 23 + 24 + 24 + 23 + 22}{7} = 22.72 = 23 \ (\text{to 2 s.f.})$$

b The results from the titration between hydrochloric acid and 25.0 cm^3 of sodium hydroxide were recorded in the table. Concordant results are those that are within ±0.10 cm^3 of each other.

Use the concordant results to calculate the mean volume of hydrochloric acid required to neutralise the 25.0 cm^3 of sodium hydroxide.

	Titration 1	Titration 2	Titration 3	Titration 4
Final burette reading/cm^3	26.10	25.20	25.45	25.15
Initial burette reading/cm^3	0.00	0.10	0.00	0.00
Volume/cm^3	26.10	25.10	25.45	25.15

Step 1 Determine the concordant results.

The results of titrations 1 and 3 are not concordant and not used to calculate the mean titre.

Step 2 Mean titre $= \dfrac{25.10 + 25.15}{2} = 25.13 \text{ cm}^3$

> **Key term**
>
> Concordant results: Concordant means exactly the same, so concordant results are the exact same readings.

(B) Guided question

1 A student carried out five titrations, adding sulfuric acid to 25.0 cm^3 sodium hydroxide solution. Her results are shown in the table.

	Titration 1	Titration 2	Titration 3	Titration 4	Titration 5
Volume of 0.100 mol/dm^3 H$_2$SO$_4$ in cm^3	26.40	26.15	27.05	27.15	27.15

Concordant results are within 0.10 cm^3 of each other. Use the student's concordant results to work out the mean volume of 0.100 mol/dm^3 sulfuric acid added.

Step 1 Examine the volumes and decide whether any are not concordant.

Titrations' 1 and 2 results are not concordant and are ignored in calculation of the mean.

Step 2 Add together the three concordant titres.

27.05 + 27.15 + 27.15 = ...

Step 3 Divide this sum by the number of concordant titres (3).

....

C Practice questions

2 A student measured the melting point of a solid four times and recorded the results in the table.

Melting point /°C	35	36	37	37

Calculate the mean melting point of the solid. Give your answer to 2 significant figures.

3 The table shows the level of nitrates present in water from two different rivers taken at four different points along the river.

River	Nitrate level mg/l				
	Point 1	Point 2	Point 3	Point 4	Point 5
A	14	13	11	9	8
B	8	9	10	11	9

Nitrate levels below 10 mg/l are considered safe to drink.

Calculate the mean nitrate level in each river. Which river is safest to drink from?

4 The results from the titration between hydrochloric acid and 25.0 cm^3 of sodium hydroxide were recorded in the table. Concordant results are those that are within ±0.10 cm^3 of each other.

   a Calculate the volume of acid used in each titration.
   b Use the concordant results to calculate the mean volume of hydrochloric acid required to neutralise the 25.0 cm^3 of sodium hydroxide.

	Titration 1	Titration 2	Titration 3	Titration 4
Initial burette reading /cm^3	0.00	14.00	0.00	15.30
Final burette reading /cm^3	13.00	26.50	12.45	28.00
Volume of HCl /cm^3				

Calculating weighted means

A weighted mean is where some values contribute more to the mean than others.

Relative atomic mass (A_r) is a weighted mean of isotopic masses. The relative atomic mass of an element can be calculated from the relative isotopic masses of the isotopes (which are the same as the mass numbers) and the relative proportions in which they occur (abundance).

$$\text{Relative atomic mass} = \frac{\text{sum of (mass} \times \text{abundance) for all isotopes}}{\text{total abundance}}$$

A Worked example

1 **Calculate the relative atomic mass of rubidium to 2 decimal places.**

	relative isotopic mass	abundance /%
^{85}Rb	85	72.15
^{87}Rb	87	27.85

Step 1 Find the total abundance.

   $72.15 + 27.85 = 100$

Step 2 Calculate the relative atomic mass.

$$\text{Relative atomic mass} = \frac{(85 \times 72.15) + (87 \times 27.85)}{100} = 85.56$$

B Guided question

1 **The element magnesium contains 79% ^{24}Mg, 10% ^{25}Mg and 11% ^{26}Mg. Calculate the relative atomic mass of magnesium to 1 decimal place.**

Step 1 First, find the total abundance.

$79 + 10 + 11 = $ ....

Step 2 Then calculate relative atomic mass by multiplying each relative isotopic mass value by its abundance and dividing by the total abundance.

$$\text{Relative atomic mass} = \frac{(79 \times 24) + (10 \times ......) + (...... \times ......)}{100} = ...$$

C Practice questions

2 The table shows the relative abundance of the two main isotopes of copper. Calculate the relative atomic mass of copper to 1 decimal place.

Isotope	^{63}Cu	^{65}Cu
Percentage abundance/%	69	31

3 The iostopes of sulfur and their abundance are shown in the table. Calculate the relative atomic mass of sulfur to 2 decimal places.

Isotope	Percentage abundance/%
^{32}S	95.02
^{33}S	0.76
^{34}S	4.22

Bar charts and histograms

Data can come in many different forms. One form is numerical and involves numbers and is called quantitative data. This can be continuous or discontinuous (discrete). Continuous data can take any value within a range such as temperature or weight and is usually rounded. Discontinuous data can only take certain values, such as the number of students – you can't have half a student!

The other type of data, qualitative data, involves words or sentences representing a particular category, like 'men' and 'women'.

Different types of graphs and charts can be drawn depending on the different types of data given.

Bar charts are used to show comparisons between categories. The *x*-axis shows discontinuous data and the *y*-axis shows continuous data. The bars are of equal width and there are gaps between the bars showing that the data is discrete and in their own categories. Sometimes, bar charts compare two or more sets of data and use a key to show what each bar represents.

Histograms are similar to a bar graph; however, the *x*-axis show continuous data. A histogram has no gaps between the bars. The area of the bar is proportional to the number of items it represents.

★ **Not explicitly required for CCEA GCSE Chemistry students.**

Key terms

Discrete data: Data which can only have particular values, such as the number of marbles in a jar.

Continuous data: Data which can take any value within a range, such as the mass of a beaker.

Bar charts: Charts showing discrete data in which the height of the unconnected bars represents the frequency.

Histograms: Charts showing continuous data in which the area of the bar represents the frequency.

A Worked example

1 **The table shows the pH of different household substances. Draw a bar chart of this data.**

Household substance	pH
shampoo	6
mouthwash	10
baking soda	9
lemon juice	3
vinegar	3
water	7

Step 1 Decide what information should go on each axis and choose a suitable scale.

The horizontal x-axis should show the different types of household substance. The y-axis should show the pH.

Step 2 For the first substance – shampoo – draw a bar extending from the x-axis up to the correct value on the y-axis. Then leave a uniform space and draw the next bar of the same width for mouthwash.

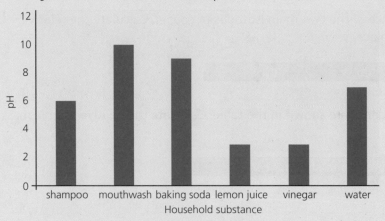

B Guided question

1 **The table shows the amounts of dissolved ions in a sample of drinking water.**

Ion	Mass in mg per dm^3
Cl^-	250
Mg^{2+}	100
NO_3^-	40

Use the information in the table to complete the bar chart.

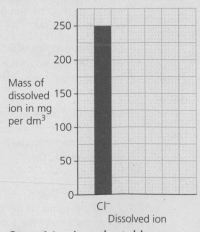

Step 1 Look at the table – you need to draw two more bars, one for Mg^{2+}, the other for NO_3^-. In a bar chart there should be a space between each bar and each bar should be the same width. On the x-axis place the labels for Mg^{2+} and for NO_3^-.

Step 2 Draw in the bar for Mg^{2+}. It should be the same width as the Cl^- bar and reach up to 100.

Step 3 Draw in the bar for NO_3^-. It should be the same width as the others and end at 40. Look carefully at the *y*-axis scale; each small box is equal to 5 mg.

C Practice questions

2 The densities at room temperature of some of the elements in Period 3 of the Periodic Table are shown below.

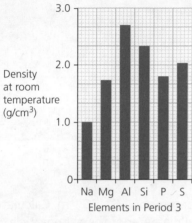

a Name the element that is most dense at room temperature.
b Name the least dense metal.
c What is the density of silicon at room temperature?
d Chlorine is in Period 3 but the density of chlorine is too small to be shown on the chart. Name one reason for this.
e State the trend in the density of metals across the period.
f Is this a bar chart or a histogram?

Order of magnitude calculations

Orders of magnitude allow us to compare very large and very small values to each other. This is useful when comparing the size of different types of particles.

An order of magnitude is a division or multiplication by 10. Each division or multiplication by ten is termed an order of magnitude. The actual length may be approximated, as it is the relative difference that is important. Orders of magnitude are easily compared using standard form.

Key term

Order of magnitude:
If we write a number in standard form, the nearest power of 10 is its order of magnitude.

A Worked example

1 **The radius of an atom is 1×10^{-10} m. The radius of the nucleus of an atom is 1×10^{-14} m. How many times bigger is the atom than the nucleus?**

You can compare the two diameters by dividing the larger power of 10 by the smaller one.

$$1 \times 10^{-10} \div 1 \times 10^{-14} = 10^4$$

So, the diameter of the atom is $10000(10^4)$ times bigger than that of the nucleus.

B Guided question

1 A sports drink contains $1.18\,\mu g/cm^3$ of potassium ions and $118\,mg/cm^3$ of sugar. Calculate how many times greater the concentration of sugar is than the concentration of potassium ions.

Step 1 Make sure both concentrations are in the same units. Convert both to g/cm^3 (or both to mg/cm^3).

potassium ions concentration:

$1.18\,\mu g/cm^3 = 0.00000118\,g$

sugar concentration:

....

Step 2 Divide the bigger concentration by the smaller.

....

C Practice questions

2 A coarse particle has a diameter of $1 \times 10^{-4}\,m$ and a nanoparticle has a diameter of $1 \times 10^{-9}\,m$. Calculate how much bigger a coarse particle is than a nanoparticle.

3 A fine particle has a diameter of $1.0 \times 10^{-6}\,m$. A nanoparticle has a diameter of $1.6 \times 10^{-9}\,m$. Calculate how many times bigger the diameter of the coarse particle is than the diameter of the nanoparticle.

» Algebra

Algebra is a branch of mathematics that uses equations in which letters represent numbers.

Understanding symbols

In Chemistry, you need to be familiar with different symbols that can be used in algebraic equations.

★ **Not explicitly required for WJEC or CCEA GCSE Chemistry students.**

Table 1.3 Commonly used symbols in Science

Symbol	Meaning
=	is equal to
>	is greater than
<	is less than
~	is similar to
∝	is proportional to

A Worked examples

1 **Place the correct symbol from the list below into the box to complete the equation** $4 \times 4 \,\square\, 12$

Step 1 Calculate $4 \times 4 = 16$

Step 2 Compare the value of 16 to the value of 12

16 is greater than 12 so choose the > symbol

$4 \times 4 > 12$

B Guided questions

1   **Place the correct symbol from the list below into the box to complete the equation 1.6×10^{-2} ☐ 0.12.**

Step 1 Write the number 1.6×10^{-2} in decimal form.

$1.6 \times 10^{-2} = 0.016$

Step 2 Compare the value 0.016 to the value 0.12 and decide if 0.016 is greater than, less than or equal to 0.12 and so choose the correct sign.

0.016 .... 0.12

C Practice questions

1   Decide whether the following are true or false.

a $10 < 3$ d $2.3 \times 10^2 = 2300$

b $15 > 8$ e $989\,000 < 9.89 \times 10^2$

c $1000 = 10^3$

Changing the subject of an equation

An equation shows that two things are equal: it will have an equals sign '='.

This means that what is on the left of the equals sign is equal to what is on the right. An example of an equation is

$$\text{moles} = \frac{\text{mass (g)}}{M_r}$$

The subject of an equation is the single variable (usually on the left of the '=') that everything else is equal to. In this example, the subject is 'moles'.

To help with chemical calculations it is very useful to be able to rearrange an equation and so change the subject of the equation. The general rule to rearrange an equation is that you must do the same to each side in order for both sides to remain equal.

A Worked example

1   a   **Make x the subject of the equation $y = x + 6$**

Step 1 Switch sides to get the new subject on the left.

$x + 6 = y$

Step 2 To get the 'x' by itself on the left-hand side, you need to subtract 6 from the left side but, to keep the equation true, we need to subtract 6 from the right side as well.

$x + 6 - 6 = y - 6$

Step 3 Simplify.

$x = y - 6$

b **Make mass the subject of this equation, moles $= \dfrac{\text{mass}}{M_r}$**

Step 1 Switch sides to get the new subject on the left.

$$\frac{\text{mass (g)}}{M_r} = \text{moles}$$

Step 2 To get mass by itself on the left-hand side, you need to remove M_r by multiplying both sides by M_r and cancelling the M_r on the left.

$$\frac{\text{mass} \times \cancel{M_r}}{\cancel{M_r}} = \text{moles} \times M_r$$

So, the answer is mass $=$ moles $\times M_r$

B Guided questions

1 **Make x the subject of the equation $y - 3 = 3x + 6$**

Step 1 Switch sides to get the new subject on the left.

$3x + 6 = y - 3$

Step 2 To get $3x$ by itself as the subject on the left-hand side, you need to subtract 6 from the left side. But, to keep the equation true, you need to subtract 6 from the right side as well and then simplify.

$3x + 6 - 6 = y - 3 - 6$

$3x = \dots\dots\dots\dots$

Step 3 To get x by itself as the subject, you need to divide the left side by 3 and to keep the equation true you must also divide the right side by 3

....

2 **Make volume the subject of the equation, moles $= \dfrac{(\text{volume} \times \text{conc.})}{1000}$**

Step 1 Switch sides to get the new subject on the left.

$$\frac{(\text{volume} \times \text{conc.})}{1000} = \text{moles}$$

Step 2 To get volume by itself as the subject on the left-hand side, you need to multiply both sides by 1000 and simplify.

$$\frac{(\text{volume} \times \text{conc.} \times \cancel{1000})}{\cancel{1000}} = \text{moles} \times 1000$$

Step 3 To get volume by itself, you now need to divide both sides by the conc.

....

C Practice questions

3 Rearrange the following equations to make x the subject.

a $y = 2x + 1$

b $3x = 4 + y$

c $x + y = 8$

d $y = mx + c$

e $3 = 4x + 2y$

f $2y + 3 = 4 - x$

4 Rearrange the equations to make the variable in bold the subject.

a $\text{moles} = \dfrac{\textbf{mass}}{M_r}$

b $\text{moles} = \dfrac{\textbf{vol}\,(\text{dm}^3)}{24}$

c $\text{percentage yield} = \dfrac{\text{actual yield} \times 100}{\textbf{theoretical yield}}$

d $\text{volume} = \dfrac{\text{moles} \times \textbf{conc.}}{1000}$

e $\text{mean rate of reaction} = \dfrac{\text{quantity of reactant used}}{\textbf{time taken}}$

Substituting into algebraic equations and solving

Key equations

Throughout your Chemistry course there are several key equations that you need to learn. These are given in Table 1.4.

Table 1.4 **Key Chemistry equations**

$\text{amount in moles} = \dfrac{\text{mass (g)}}{M_r}$	where M_r = relative formula mass
$\text{amount in moles} = \dfrac{\text{mass (g)}}{A_r}$	where A_r = relative atomic mass
$\text{percentage yield} = \dfrac{\text{actual yield}}{\text{theoretical yield}} \times 100$	
$\text{atom economy} = \dfrac{\text{sum of relative formula mass of desired product from equation}}{\text{sum of relative masses of all reactants from equation}} \times 100$	
$\text{amount in moles of a gas} = \dfrac{\text{volume (dm}^3)}{24}$	
$\text{amount in moles} = \dfrac{\text{volume (cm}^3) \times \text{conc. (mol/dm}^3)}{1000}$	
$\text{mean rate of reaction} = \dfrac{\text{quantity of reactant used}}{\text{time taken}}$	
$R_f = \dfrac{\text{distance moved by substance}}{\text{distance moved by solvent}}$	

When solving a problem using these equations, it is important that you change the subject if necessary and then substitute the numerical values given in the question. You must use the correct units for measurement in an equation when substituting numerical values into it.

A Worked example

1 **Calculate the amount, in moles, of calcium hydroxide present in 25.0 cm³ of a solution of concentration 0.25 mol/dm³.**

Step 1 The question gives a volume and a concentration and the number of moles is to be calculated. Hence, the equation to use is:

$$\text{amount in moles} = \frac{\text{volume (cm}^3) \times \text{conc. (mol/dm}^3)}{1000}$$

Step 2 The subject is amount in moles, so does not need to be changed. Simply substitute the numerical values and calculate the answer.

$$\text{amount in moles} = \frac{\text{volume (cm}^3) \times \text{conc. (mol/dm}^3)}{1000}$$

$$= \frac{25.0 \times 0.25}{1000} = 0.0063 \text{ to 2 significant figures}$$

> **Tip**
> . . . .
> If the volume of calcium hydroxide was given in dm³ then you use the equation moles = volume (cm³) × conc. (mol/dm³) and do not divide by 1000.

B Guided question

1 **Calculate the concentration of calcium hydroxide solution obtained when 0.0034 moles of calcium hydroxide is dissolved in 15.0 cm³ of water.**

Step 1 The information in the question is moles=0.0034 mol and volume=15.0 cm³. Hence, use the equation:

$$\text{amount in moles} = \frac{\text{volume (cm}^3) \times \text{conc. (mol/dm}^3)}{1000}$$

Step 2 The subject is amount in moles, but you need to find the concentration, so the subject needs to be changed to conc.

Switch the sides so that conc. is on the left.

$$\frac{\text{volume (cm}^3) \times \text{conc. (mol/dm}^3)}{1000} = \text{amount in moles}$$

Step 3 You require conc. on its own on the left. To remove the 1000, multiply both sides by 1000 and simplify.

$$\frac{\text{volume} \times \text{conc.} \times \cancel{1000}}{\cancel{1000}} = \text{moles} \times 1000$$

....

Step 4 To get conc. on its own on the left, divide both sides by volume and simplify.

....

Step 5 Then substitute the numerical values – moles=0.0034 mol and volume=15.0 cm³ and calculate the answer.

....

C Practice questions

2 In a chromatography experiment the distance moved by the solvent is 10.2 cm. Calculate the distance moved by a substance if its R_f value is 0.80.

3 Calculate the percentage atom economy for making copper(II) sulfate from copper carbonate and sulfuric acid.

$$CuCO_3 + H_2SO_4 \rightarrow CuSO_4 + H_2O + CO_2$$

Relative formula masses: $CuCO_3 = 124$, $H_2SO_4 = 98$, $CuSO_4 = 160$, $H_2O = 18$, $CO_2 = 44$

4   Calculate the mass in 0.45 mol of potassium nitrate. Give your answer to 2 significant figures.

5   In a reaction 0.20 g of magnesium was completely reacted with sulfuric acid in 30 seconds. Calculate the mean rate of reaction during this time in g/s of magnesium used.

» Graphs

Translating information between graphical and numeric form

Plotting graphs

Experiments often involve variables. The independent variable is the factor that is being changed during the experiment. The dependent variable is the factor that is being measured during the experiment. The controlled variables are the factors that are kept constant during the experiment. Changing the independent variable causes a change in the dependent variable.

After carrying out an experiment, it is often useful to plot a graph to help analyse your results. A graph is an illustration of how two variables relate to one another. In Chemistry you are often asked to draw a line graph with a best-fit curve or line.

General construction

When plotting a graph, it is important to pay attention to the following points.

●   Choose a suitable scale to make sure you use as much of the graph paper as possible – at the very least, half of the graph paper should be used. The two graphs in Figure 1.7 illustrate this.

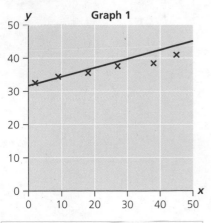
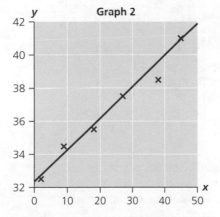

| Graph 1 – a poor scale in the y direction which compresses the points into a small section of graph paper.   ✗ | Graph 2 – a good scale as the points fill more than half the graph paper in both x and y directions   ✔ |

▲ Figure 1.7 Choosing scales

●   It is also important when choosing a scale that you examine the data to establish whether it is necessary to start the scale(s) at zero.
●   Choose a simple scale increasing in multiples of 2, 5 or 10 – avoid using multiples of 3 or 7.

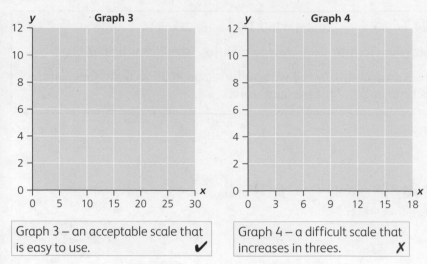

Graph 3 – an acceptable scale that is easy to use. ✔

Graph 4 – a difficult scale that increases in threes. ✗

▲ Figure 1.8 Choosing scales

- The independent variable is placed on the *x*-axis and the dependent variable on the *y*-axis. Axes should be labelled with the name of the variables and the unit of measurement for each. For example, one of the labels may be 'temperature/°C' or 'temperature in °C'.
- Data points should be marked with a cross (x) so that all points can be seen when a line of best fit is drawn.
- A line of best fit should be drawn. When judging the position of the line, there should be approximately the same number of data points on each side of the line; resist the temptation to simply connect the first and last points. The line of best fit can be either a straight line or a curve.

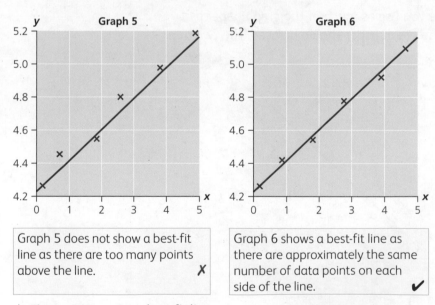

Graph 5 does not show a best-fit line as there are too many points above the line. ✗

Graph 6 shows a best-fit line as there are approximately the same number of data points on each side of the line. ✔

▲ Figure 1.9 Drawing a best-fit line

Tip
A line of best fit is added by eye. You should use a transparent plastic ruler or a flexible curve to aid you.

Not all lines of best fit go through the origin – before using the origin as a point always ask the question 'Does a 0 in the independent produce a 0 in the dependent?'

- When drawing a best-fit line or curve, ignore any anomalous results.

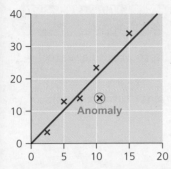

▲ Figure 1.10 Ignore anomalies when drawing a best-fit line

- The graph should have a title that summarises the relationship that is being illustrated – this should include the independent variable, the dependent variable and the reaction being studied. For example, a suitable title is 'A graph of concentration against time for the reaction between magnesium and hydrochloric acid' or simply, 'A concentration–time graph for the reaction between magnesium and hydrochloric acid'.

A Worked examples

1 In an experiment, a lump of calcium carbonate was added to 50 cm^3 of hydrochloric acid in a conical flask and placed on a balance. A stopwatch was started as soon as the calcium carbonate made contact with the acid, and the mass was recorded every 20 seconds in the table. Plot a graph of mass (y-axis) against time (x-axis).

Time in s	0	20	40	60	80	100	120
Mass in g	234.10	237.70	233.40	233.20	233.05	233.00	233.00

Step 1 Decide on the scale for the x-axis. The graph paper has 12 squares across, hence it is appropriate to start at time zero and increase in intervals of 10 up to 120 seconds.

Step 2 Decide on the scale for the y-axis. The y-axis has 16 squares up. The masses range from 234.10 to 233.00, which is an interval of 1.1 g. It is not sensible to start this scale at zero as this would not spread out the data points over the graph paper. Instead, begin at 232.80 and each square could represent 0.10 g. Count the number of zeros after the 1.

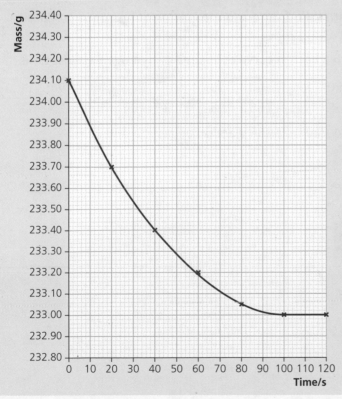

B Guided questions

1 Hydrogen peroxide decomposes in the presence of solid manganese(IV) oxide to produce water and oxygen.

Use the data in the table to draw a graph of mass of oxygen lost (y-axis) against time (x-axis).

Time/min	1	2	3	4	5	6	7
Mass of oxygen lost/g	0.20	0.34	0.38	0.45	0.47	0.48	0.48

Step 1 Decide on the scale for the x-axis. The graph paper has 7 squares across, hence it is appropriate to start at time zero and increase in intervals of 1 up to 7 minutes.

Step 2 Decide on the scale for the y-axis. The graph paper has 6 squares up, hence it is appropriate to start at mass 0.00g and increase in intervals of 0.10 up to 0.50g.

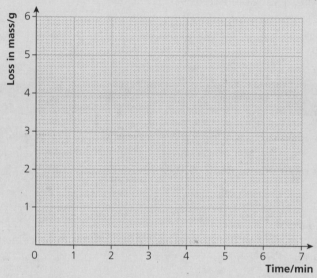

C Practice questions

1 In an experiment, magnesium was reacted with sulfuric acid and the volume of hydrogen produced collected and measured, every 10 seconds, in a gas syringe.

Use the data in the table to draw a graph with a best-fit curve of volume of hydrogen against time.

Time /s	0	10	20	30	40	50	60	70	80	90	100
Volume of hydrogen /cm³	0	30	55	75	88	98	102	104	104	104	104

2 During an equilibrium reaction, a gas C is produced from the reaction of gases A and B.

$$A(g) + B(g) \rightleftharpoons C(g)$$

The percentage of C in the reaction mixture varies with temperature. Plot a graph of percentage of C against temperature.

Temperature /°C	100	200	300	400	500
Percentage of C in equilibrium mixture /%	58	42	30	21	16

3 In an experiment, some calcium carbonate and acid were placed in a conical flask on a balance and the balance reading recorded every minute. The results were recorded and the graph shown was drawn.

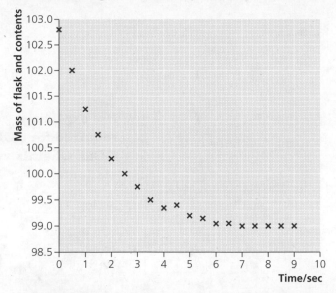

a Are there any results that you would ignore when drawing a best-fit curve?
b Look at the labels on the axis and describe any changes you would make.
c Suggest a title for this graph.
d Do you think the scale is appropriate in this graph? Explain your answer.

Determining the slope and intercept of a linear graph of equation $y = mx+c$

Every straight-line linear graph can be represented by an equation: $y = mx + c$. The co-ordinates of every point on the line will solve the equation if you substitute them in the equation for x and y. The equation also contains two constants, m and c. The constant c is called the intercept (in this case the y-intercept) and m is the gradient.

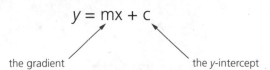

the gradient   the y-intercept

▲ Figure 1.11 The equation for a straight line

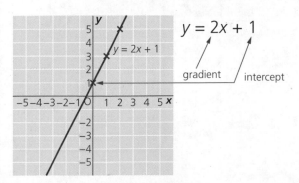

▲ Figure 1.12 Straight-line graph

★ **Understanding that $y = mx + c$ represents a linear relationship is not explicitly required for WJEC or CCEA GCSE Chemistry students.**

Key terms

Intercept: This is the point where the graph crosses an axis. In the equation $y = mx+c$, c is the y-intercept where the graph crosses the y-axis when $x=0$, it is the value for y when $x=0$.

Gradient: This is another word for 'slope'. It is the change in the y-value divided by the change in the x-value.

Gradient is another word for 'slope'. The higher the gradient of a graph at a point, the steeper the line is at that point. A positive gradient means the line slopes up from left to right. A negative gradient means that the line slopes down from left to right. For a straight-line graph, the gradient is a constant value. A zero gradient graph is a horizontal line.

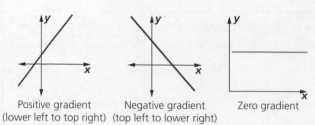

Positive gradient
(lower left to top right)

Negative gradient
(top left to lower right)

Zero gradient

▲ **Figure 1.13** Gradients

To find the gradient (m) of a straight-line graph

★ **Not explicitly required for CCEA GCSE Chemistry students.**

Choose any two points on the best-fit line and draw a triangle with the best-fit line forming the hypotenuse (longest side).

A more accurate answer for the gradient is obtained when the points are as far apart as possible.

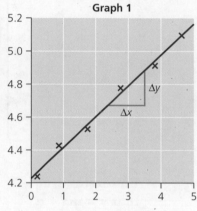

Graph 1

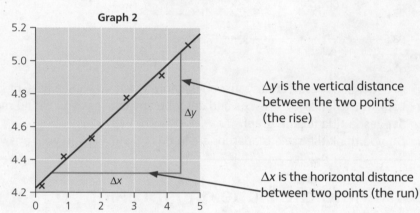

Graph 2

Δy is the vertical distance between the two points (the rise)

Δx is the horizontal distance between two points (the run)

Graph 1 – the points chosen are not far enough apart and the hypotenuse (red line) is too small to use to calculate the gradient accurately. ✗

Graph 2 – the triangle is a much better size, the hypotenuse (red line) is more than half the length of the best-fit line. ✔

▲ **Figure 1.14** Finding a gradient

To calculate the gradient, use the equation

$$\text{gradient (m)} = \frac{\text{change in } y\text{-axis}}{\text{change in } x\text{-axis}} = \frac{\Delta y}{\Delta x}$$

> **Tip**
>
> It may be easier to remember gradient
>
> $\text{gradient} = \dfrac{\text{rise}}{\text{run}}$

> **Tip**
>
> You should always show your working when calculating the gradient of a line. It is also a good idea to show and label the triangle used.

(A) Worked example

1 Calculate the gradient of the line shown.

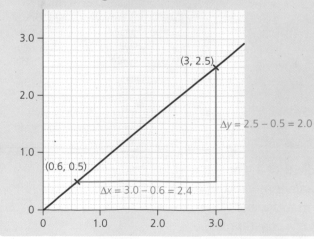

(3, 2.5)

$\Delta y = 2.5 - 0.5 = 2.0$

(0.6, 0.5)

$\Delta x = 3.0 - 0.6 = 2.4$

$$\text{gradient } (m) = \frac{\text{change in } y\text{-axis}}{\text{change in } x\text{-axis}} = \frac{\Delta y}{\Delta x}$$

$$m = \frac{2.5 - 0.5}{3.0 - 0.6} = \frac{2.0}{2.4} = 0.8$$

B Guided question

1 **a** **Find the gradient of the lines in A, B and C.**

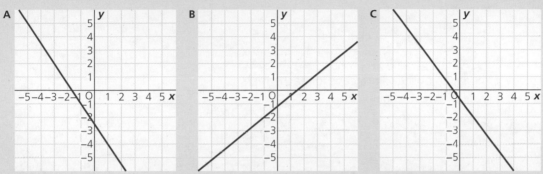

Step 1 Choose two points that are far apart on the line; these will form the hypotenuse of the triangle. This step has been completed for each graph below and the points are marked as crosses.

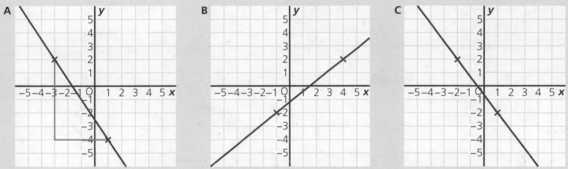

Step 2 Complete the triangle – these are the green lines that have been completed in Graph A.

Step 3 Find the Δy (rise) value.

....

Step 4 Find the Δx (run) value.

....

Step 5 Find the gradient using the equation:

$$\text{gradient } (m) = \frac{\text{change in } y\text{-axis}}{\text{change in } x\text{-axis}} = \frac{\Delta y}{\Delta x}$$

....

Step 6 Decide whether it is a positive gradient sloping from lower left to top right or a negative gradient.

....

b **Find the y-intercept and write the equation for the line.**

Step 1 Write down the number at which the blue line cuts through the y-axis (at $x = 0$). This is the intercept, c.

....

Step 2 Substitute the values for m and for c into the equation $y = mx + c$.

....

C Practice questions

2 For each of the graphs A to G, decide whether the gradient is positive, negative or zero.

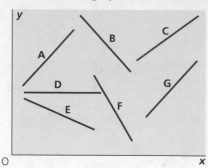

3 a Work out the gradient of each graph A to F shown below.
b Write down the value of the *y*-intercept for each graph.
c Write the equation for each line.

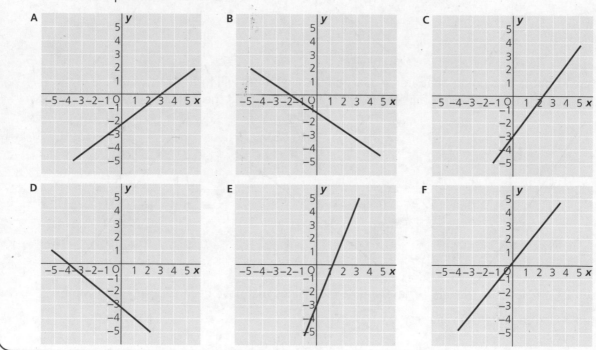

Drawing and using the slope of a tangent to a curve

★ **Not explicitly required for CCEA GCSE Chemistry students.**

The word **tangent** means 'touching' in Latin.

To draw a tangent at a point (x, y), follow these steps:

1 Place your ruler through the point (x, y).

2 Make sure your ruler goes through the point and does not touch the curve at any other point.

3 Draw a ruled pencil line passing through point (x, y).

Key term

Tangent: This is a straight line that just touches the curve at a given point and does not cross the curve.

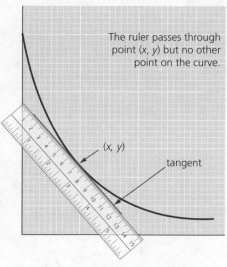

▲ Figure 1.15 Tangent to a curve

The ruler passes through point (x, y) but no other point on the curve.

(x, y)

tangent

To calculate the gradient of a curve at a particular point, it is necessary to draw a tangent to the curve at the point and calculate the gradient of the tangent.

A Worked example

1 **In an experiment a student recorded the total volume of gas collected in a reaction of 20-second intervals.**

Time /s	0	20	40	60	80	100	120
Volume of gas /cm³	0	21	42	56	65	72	72

a) **Plot a graph using the data shown and draw a line of best fit.**

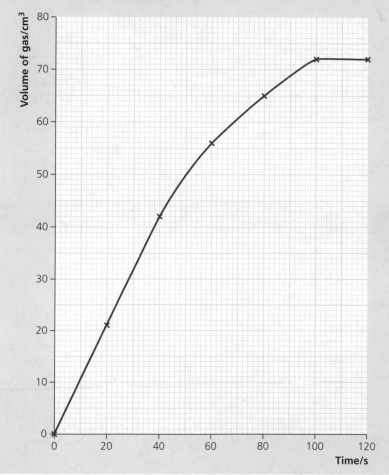

The graph will be a graph of volume of gas (*y*-axis) against time (*x*-axis). Allocating one large square for 20 s is a scale that is suitable on the *x*-axis, and one large square for 10 cm^3 of gas is suitable on the *y*-axis.

b) **Use the graph to calculate the rate of reaction at 60 s in cm^3/s.**

The graph is a curve and to find the rate of reaction at 60 s, a tangent to the curve must be drawn at 60 s as shown in the folllowing graph in red.

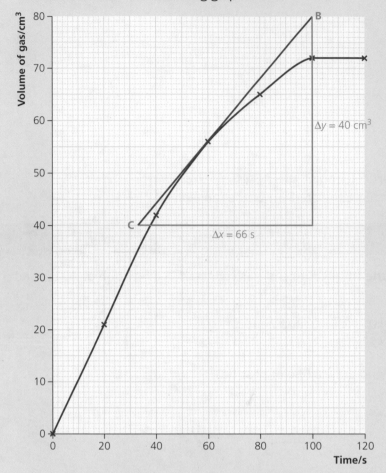

The rate is given by the gradient of this tangent. To find the gradient, choose two points B and C that are far apart on the line and form a triangle, as shown in green on Figure 1.29.

$$\text{gradient } (m) = \frac{\text{change in } y\text{-axis}}{\text{change in } x\text{-axis}} = \frac{\Delta y}{\Delta x} = \frac{40}{66} = 0.61 \, \text{cm}^3/\text{s} \ (\text{to 2 s.f.})$$

B Guided question

1 What is the gradient of the curve at point A?

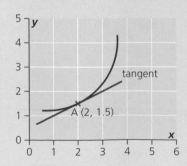

Step 1 To find the gradient of the curve at point A, a tangent to the curve at point A must be drawn. It is shown in red in the graph above.

Step 2 Choose two points, B and C, far apart on the line and form a triangle as shown in the graph below.

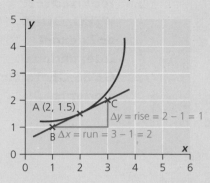

A (2, 1.5)

C

Δy = rise = 2 − 1 = 1

B Δx = run = 3 − 1 = 2

Step 3 Calculate the gradient using:

$$\text{gradient (m)} = \frac{\text{change in } y\text{-axis}}{\text{change in } x\text{-axis}} = \frac{\Delta y}{\Delta x}$$

C Practice questions

2 Calculate the rate of reaction when the concentration of an ester is 0.200 mol/dm³.

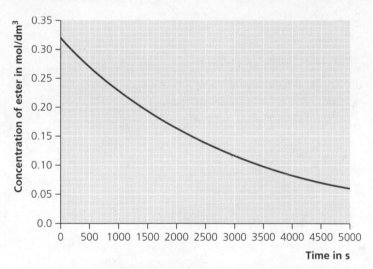

Concentration of ester in mol/dm³

Time in s

3 The volume of carbon dioxide gas produced over time when calcium carbonate reacted with hydrochloric acid was recorded in the table.

Time /s	0	10	20	30	40	50	60	70	80	90	100
Volume of carbon dioxide /cm³	0	22	35	43	48	52	55	57	58	58	58

a Plot a graph of the volume of carbon dioxide against time.
b Calculate the rate of reaction at 20 seconds by drawing a tangent to the curve.
c Calculate the rate of reaction at 60 seconds by drawing a tangent to the curve.

>> Geometry and trigonometry

Representing 2D and 3D

Table 1.5 shows the difference between 2D and 3D shapes.

★ Not explicitly required for WJEC, CCEA or Edexcel International GCSE Chemistry students.

Table 1.5 Comparison of 2D and 3D shapes

3D shapes	2D shapes
Have three dimensions – height, depth and width.	Have two dimensions – length and width. They have no depth and are flat.
These figures can be drawn on a sheet of paper using wedged and dashed lines.	These figures can be drawn on a sheet of paper in one plane using solid lines.
3D figures deal with three co-ordinates: the x-co-ordinate, y-co-ordinate and z-co-ordinate.	2D figures deal with two co-ordinates: the x-co-ordinate and y-co-ordinate.

Displayed formulae and structural formulae represent molecules in 2D with all the covalent bonds shown as solid lines. No information about the orientation or shape of the molecules is given.

The displayed structural formula of methane and ethanol are shown in Figure 1.16.

▲ Figure 1.16 Methane and ethanol

When drawing 2D structures it usually does not matter which angle you draw the atoms at. This is because molecules are three-dimensional so when representing them in two dimensions on paper, no diagram will represent them as they truly are. For example, each of the four structures in Figure 1.17 is a correct displayed formula of propene (C_3H_6).

▲ Figure 1.17

When molecules are drawn in 3D the symbols used are those shown in Table 1.6.

Table 1.6 Types of bond

Type of bond	Orientation
Normal bond ———	Bond lies in the plane of the paper
Dashed bond --------	Bond extends backwards, effectively into the page
Wedged bond ◄▬	Bond extends forwards, effectively out of the page

A 3D structural representation of methane CH_4, with its ball and stick model is shown in Figure 1.18.

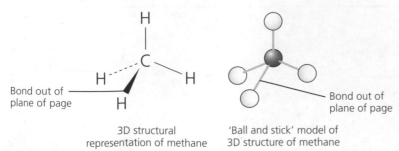

Bond out of plane of page

3D structural representation of methane

'Ball and stick' model of 3D structure of methane

Bond out of plane of page

▲ Figure 1.18 3D ball and stick model of methane (CH_4)

Graphite and diamond are giant covalent molecules. You should practice drawing diagrams of graphite and diamond.

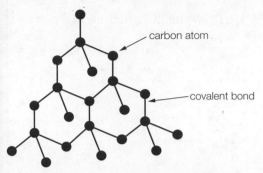

▲ **Figure 1.19** Structure of diamond

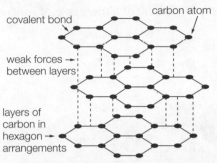

▲ **Figure 1.20** Structure of graphite

Ⓐ Worked example

1 **A molecule of ethanol has eight single covalent bonds. Draw the missing bonds in the diagram below to complete the displayed structural formula of ethanol.**

$$
\begin{array}{c}
\text{H} \\
| \\
\text{H}-\text{C}-\text{C} \\
| \\
\text{H}
\end{array}
$$

Step 1 Ethanol is C_2H_5OH. First, fill in the bonds between the second carbon and two hydrogens as shown in red below.

$$
\begin{array}{cc}
\text{H} & \text{H} \\
| & | \\
\text{H}-\text{C}-&\text{C} \\
| & | \\
\text{H} & \text{H}
\end{array}
$$

Step 2 Ethanol contains the OH group. Remember that there is a bond between the O and the H that should be shown. Draw a bond between the carbon and the O, and then another bond to the H (shown in red).

$$
\begin{array}{cc}
\text{H} & \text{H} \\
| & | \\
\text{H}-\text{C}-&\text{C}-\text{O}-\text{H} \\
| & | \\
\text{H} & \text{H}
\end{array}
$$

Ⓑ Guided question

1 **The diagram below shows a 3D ball and stick model of a molecule of ammonia. Draw the 2D structure of ammonia.**

Step 1 You need to remember that ammonia has the formula NH_3 and so the red ball represents a nitrogen atom and the other three balls are hydrogen atoms.

Step 2 When drawing a 2D structure from a 3D representation, remember that each stick represents a covalent bond, which is drawn as a single line. First, draw the central nitrogen atom, as shown, then complete by drawing two more bonds to two hydrogen atoms.

N — H

C Practice questions

2 The figure below shows a 3D model of a molecule of methane (CH$_4$). Draw the 2D structure of a methane molecule.

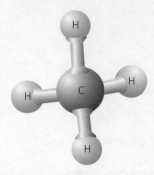

3 A molecule of methanol has five single covalent bonds. Draw the missing bonds in the following diagram to complete the displayed structural formula of methanol.

H — C

4 The figure below shows a dot and cross diagram of a molecule of water. Draw the 2D structure of water representing each covalent bond with a single line.

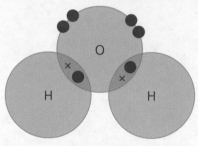

5 Draw the displayed structural formula of

  a ethane

  b propene

  c but-1-ene.

Area, surface area and volume

It is possible to calculate the area of triangles and rectangles using the formula:

$$\text{Area of a triangle} = \frac{1}{2} \times \text{base} \times \text{height}$$

$$\text{Area of rectangle} = \text{length} \times \text{height}$$

The units of area are often mm^2, cm^2 or m^2.

You should also be able to calculate the surface area and volume of a cube. Surface area is the area of all the surfaces of a 3D shape added together. Volume is measure in cubed units such as mm^3, cm^3 or m^3.

It is useful to work out the surface area to volume ratio, particularly to help understand rates of reaction. As the side of a cube decreases by a factor of 10 the surface area to volume ratio increases by a factor of 10 (see Practice question 4).

★ **Not explicitly required for CCEA, OCR 21st Century, or Edexcel International GCSE Chemistry students.**

A Worked example

1 a **A cube has a side length 5 cm. Calculate the surface area and volume of this cube.**

Step 1 Calculate the surface area of one face of the cube.

Surface area of one face = $5 \times 5 = 25\,cm^2$

Step 2 Calculate the surface area by multiplying by the number of faces – there are six faces.

Surface area of one cube = area of one face × number of faces
$$= 25 \times 6 = 150\,cm^2$$

Step 3 Calculate the volume of the cube.

Volume of cube = length × width × height
$$= 5 \times 5 \times 5 = 125\,cm^3$$

b **What is the surface area to volume ratio of this cube?**

Step 1 Write down the ratio.

surface area : volume
   150   :   125

Step 2 Simplify the ratio by dividing by a common factor such as 5. You may need to keep dividing both numbers by this factor until they can no longer be divided to give whole numbers.

Surface area	:	volume
150	:	125
30	:	25
6	:	5

The surface area to volume ratio is 6:5

Tip

You might find it helpful to turn to page 14 and recap on ratios before starting Part b. Ratios are represented using colons ':'.

B Guided question

1 **An aluminium cube has a volume of 8 cm² and a total surface area of 24 m³. Calculate the surface area to volume ratio of this cube.**

Step 1 Write down the ratio in the correct form in words and place the numbers underneath.

Surface area   :   volume
   24   :   ....

Step 2 Simplify the ratio by dividing by a factor.

C Practice questions

2 Which of the following surface area to volume ratios is the largest?

A 2:3
B 8:15
C 2:7

3 The length of one side of a cube of silver is 2 cm. Calculate the surface area to volume ratio of this cube.

4 Cube A has a side length of 5 cm and cube B has a side length of 3 cm. Which cube has the greater surface area to volume ratio?

5 Show that the surface area to volume ratio of a cube with a side length of 2 cm is ten times greater than that of a cube with a side length of 20 cm.

2 Literacy

>> How to write extended responses

Some questions on your Chemistry GCSE exam papers are extended response questions. These questions provide an opportunity for you to demonstrate your ability to construct and develop a sustained line of reasoning. In addition to assessing your knowledge or understanding of a Chemistry topic, they will also assess the quality of your written communication – in other words, your ability to write a logical answer with all points arranged in a sensible order.

Extended response questions are usually six-mark questions (although they can be anywhere between four and six) and may be indicated on your examination paper. Some boards insert a phrase similar to 'in this question part, quality of written communication will be assessed' or 'QER', which stands for 'quality of extended response', or the question may be simply marked with an asterisk (*).

Approaching an extended response question

When answering extended response questions, you need to:

- ensure that your writing is legible, and spelling, punctuation and grammar are accurate,
- organise your points clearly and logically using specialist scientific vocabulary where appropriate,
- ensure that the information you give is relevant.

Before starting your answer, it is useful to read the question carefully and ask yourself the following question.

What is the question asking?

In questions, command words are those words that tell you what the question is asking and what you need to do. These are words such as Evaluate, Describe, Explain and Compare. The command words that you need to be familiar with are given in the appendix on page 123.

After reading the extended response question, the first thing you should do is underline the command word and think about what it means because it will tell you what the question is asking. After underlining the command word, read the question again and circle any words that tell you the topic that is being tested and any other key terms.

For example, to answer the question 'Compare the isotopes of Lithium ^{6}Li and ^{7}Li', you would:

- Underline the command word 'Compare' and work out what this command word means. In this case, it wants you to describe similarities and differences.
- Then you should circle the words 'isotope' as this is the *topic* being tested, you would also circle the *key terms* '^{6}Li and ^{7}Li', which tell you which specific examples need to be included.

'<u>Compare</u> the (isotopes) of Lithium (⁶Li) and (⁷Li).'

Having thought about what the question is asking, you should then plan your answer.

How to plan your answer

If you do not plan your answer, there is a temptation to write down everything you know about the topic, which means you will include many irrelevant details. Remember that the *quality* of your response and how well you answer the question is being assessed in this style of question, not how much you can write. It is wise, therefore, to take the time to plan your answer rather than rushing straight into it.

First, think about the topic as a whole and decide what parts of your knowledge are relevant to the question. Then, consider how to structure your answer by putting the relevant points in a logical order.

For example, to answer the question 'Compare the isotopes of Lithium ^{6}Li and ^{7}Li', the command word tells you to 'Compare'. You would do this by describing the similarities and differences of the specific examples of lithium given in the question. You could plan an answer in various ways, but three possibilities are outlined here.

1 Using a table

The command word is 'Compare', so a simple table showing the similarities and differences of the lithium isotopes may be useful.

Table 2.1

Same	^{6}Li	^{7}Li	Different	^{6}Li	^{7}Li
Number of protons	3	3	Number of neutrons	3	4
Atomic number	3	3	Mass number	6	7
Number of electrons	3	3			
Electronic configuration	2,1	2,1			
Reactions (same number of electrons in outer shell)					

You can decide on the order you will cover your ideas – numbering the points in your table may help do this. Then, when you write your answer, crossing off each idea in the table once you have written about it will help you keep track.

2 Using a diagram

For this method, write down the key words you need to include and then add details about lithium, as shown in Figure 2.1. Circle each key word and classify it as a similarity or a difference.

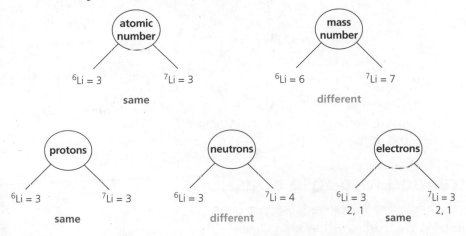

▲ Figure 2.1 Using a diagram to plan an extended response answer

Then, number the key words to show the order in which you will cover each idea. When you write your answer, cross out each key word after you have written about it.

3 Using bullet points

This method involves making short notes of what you will cover in your answer. The list format means it's easy to check through it quickly and you can also number the order you will cover each point. For example, for the isotopes question on page 48, you may write the following bullets:

- Things that are the same:
  - atomic number (3)
  - protons (3)
  - one electron in the outer shell
  - and will react in the same way.
- Things that are different:
  - mass number (6 and 7)
  - neutrons (3 and 4).

When writing your answer, you will need to make sure these notes are rewritten as continuous prose, linking your ideas in a logical way.

How to check your answer

For extended response questions you should read through your answer to check that:

- you have answered what the command word in the question asked for,
- you have used the examples or information you were asked to use,
- your spelling, punctuation and grammar is correct.

In extended response questions, correct spelling of key scientific words is important. There is a huge number of specific chemical words in an exam board specification, but it is a good idea to learn to spell the ones given in Table 2.2 – they are those that are most commonly spelt incorrectly.

Table 2.2 Correct spelling of some important chemical words.

activation energy	alkali	burette
catalyst	collision	corrosion
covalent	crystal	delocalised
distillation	hydrochloric	molecule
neutralisation	neutrons	nucleus
phytomining	pipette	polymerisation
precipitate	solution	sulfuric

Punctuation is also important. You should ensure that you use capital letters at the beginning of each sentence and a full stop at the end.

How to do well in extended response questions

To fully understand how to successfully answer an extended response question it is useful to look at how such questions are marked – they are marked differently from other questions on an exam paper and you will note that mark

> **Tip**
> In extended response questions you should only use bullet points for your plan. Use complete sentences when writing out your actual answer.

> **Tip**
> If you are asked to write an extended response answer, remember the steps using this acronym: CPWC – **command word, plan, write, check**.

schemes for these questions have three levels. This is why extended response questions are also called level of response questions.

When an examiner marks these question types, first they will read your answer as a whole and compare the quality of the content and writing with the level descriptor given on the mark scheme. They then decide which level best describes your answer – for example, if it is a strong answer they will give it a Level 3, while a weaker answer may get a Level 1. Have a quick look at pages 53 and 54 for examples of levelled mark schemes.

To decide which mark within the level is given, the examiner then looks at the indicative Chemistry content points, which are a guide to what should be included in the answer. Indicative content is the factual points that could be included to answer the question.

If the answer only barely meets the requirements the lower mark is awarded. You do not have to have all the indicative content points present to obtain full marks, but full marks are only awarded if there are no incorrect statements that contradict a correct response. The peer-response-style questions on page 56 and 56 show some example mark schemes and help you further understand how these questions are marked.

> **Tip**
>
> Remember, you do not need to have all the indicative content points to obtain full marks, but your answer does need to be factually correct.

❯❯ How to answer different command words

Let's work through the following extended writing questions, which use different command words, to help you further understand how to write a good extended answer.

For each command word there is an 'expert commentary' question that gives a sample student response, along with an analysis of what is good and bad about it. You will then be given the chance to apply what you've learnt to mark a sample answer yourself – peer assessment. Finally, you will be asked to write an answer yourself by improving on another student response in order to get the full marks.

Extended responses: 'Describe'

A question requiring you to 'describe' something is asking for a detailed written account on the relevant facts and features relating to the topic being examined. Remember that describe does not mean explain, which is a higher-level command word. You do not need to focus on causes and reasons in 'describe' questions.

A Expert commentary

1 Describe an experiment to prepare pure, dry crystals of zinc sulfate by reacting a suitable zinc compound with a suitable acid. [6]

Student answer

Measure out 25 cm³ of sulfuric acid in a measuring cylinder and place in a beaker. Warm the acid and add some spoonfuls of zinc carbonate to the acid and you will see bubbles and the zinc carbonate disappears. Keep going adding the zinc carbonate until there is some solid on the bottom of the beaker. Now separate the excess solid from the solution and place the solution in an evaporating basin. Heat the solution in the evaporating basin using a Bunsen burner until all the water has evaporated. This will leave you with pure, dry crystals of zinc sulfate.

This is a Level 2 answer worth 4 marks.

The student has identified the correct acid and stated how to measure it out.

Note that zinc oxide and zinc hydroxide could be used instead of zinc carbonate, but the answer should then state that no bubbles will be observed.

The student has mentioned the correct apparatus to hold the solution at this point: an evaporating basin.

Heating the solution until all the water has evaporated will leave powder, not crystals, so this is an error. It is better to heat to half the volume and allow to cool and crystallise.

This is incorrect terminology. Chemists should use spatulas to lift solids, not spoons.

The actual method of separating by filtering should be mentioned.

A reference to drying the crystals between sheets of filter paper should be included

B Peer assessment

2 In an experiment to prepare pure, dry crystals of potassium nitrate, a student placed 25.0 cm³ of potassium hydroxide in a conical flask and titrated it with nitric acid.

Describe the complete experimental procedure the student used to set up and carry out a titration to prepare pure dry crystals of potassium nitrate. You should name the apparatus used in the experiment in your answer. [6]

Student answer

First, use a pipette to measure out 25 cm³ of potassium hydroxide and put it into a conical flask on top of a white tile. Then add a few drops of universal indicator and add the acid from a burette, swirling the flask, until the indicator changes colour from blue to red. Then pour the solution into an evaporating basin and use a Bunsen burner to evaporate off the water. Dry the crystals between filter paper.

Use the mark scheme and indicative content to award this answer a level and a mark.

Mark scheme

Level descriptors	Marks
Level 3 A detailed, accurate and coherent description is given, which demonstrates a good knowledge and understanding of all aspects of the titration, including how pure, dry crystals are produced. It is logically presented.	5–6
Level 2 A description is given that demonstrates a reasonable knowledge and understanding of some elements of the titration and preparation of crystals. There may be some inaccuracies and some steps may be missing. The description is mostly coherent and logical	3–4
Level 1 A description is given that is not logically ordered and with significant gaps.	1–2
No relevant content	0
Indicative content	

- rinse pipette with potassium hydroxide
- measure 25.0 cm³ potassium hydroxide using a pipette
- place potassium hydroxide into a conical flask
- add some drops of a named indicator, e.g. methyl orange, phenolphthalein
- rinse burette with nitric acid
- fill burette with nitric acid and record burette reading
- add acid from burette to the flask slowly while swirling until the indicator changes colour
- state correct indicator change
- read volume of acid in burette at end of the titration
- repeat the experiment without indicator, adding the same volume of acid
- pour solution into an evaporating basin and heat until the volume reduces by half
- cool and crystallise
- dry crystals between pieces of filter paper

I would give this a level of .... and a mark of ....

This is because

....

....

....

....

Tip

• •

To assign a level, first look at the answer and decide if it is a logical series of steps that could be used to prepare the crystal. Are there any significant steps missing?

C Improve the answer

3 Describe an experiment that could be used to electrolyse molten zinc chloride. In your answer name the apparatus you would use and state any observations. [6]

Student answer

I would weigh out 10 g of zinc chloride and record the mass. I would put it in an evaporating basin. I would put two electrodes in the zinc chloride and attach them to a power pack. When electricity is switched on at one electrode there should be a grey substance and at the other a gas would be observed. A fume cupboard should be used.

Rewrite this answer to improve it and obtain the full 6 marks.

Extended responses: 'Explain'

'Explain' means to state the reasons why something happens. The points in the answer must be logically linked; for example, in this question the reason why each substance conducts should be given by naming the charge carriers.

A Expert commentary

1 Explain if the following substances conduct electricity by referring to the structures of the substances. [6]
- copper metal
- copper chloride solution
- chlorine gas.

Student answer

Copper is a metal and is used in wires and in plumbing in most households. It has lots of delocalised electrons which can move all around the layers and so it conducts electricity. This is why it is a good conductor. Copper chloride also has a metal in it so it has delocalised electrons. It cannot conduct when it is solid but when it is dissolved in solution then the delocalised electrons can move and carry the charge. Chlorine is a molecule and it does not have any plus or minus. Because it does not have a charge it cannot conduct electricity.

This is a Level 3 answer worth 5 marks.

This is a correct description of how metals conduct.

The wrong particle for conduction is identified but the idea that particles can only move to conduct when in solution (or molten) is correct.

This is extra information that is irrelevant, and a waste of time to include. It may also lead to inaccuracies in your answer.

Copper metal is not present in copper chloride so there are no delocalised electrons. Copper chloride is made up of copper ions and chloride ions that move and carry charge.

The student correctly recognises that charges are needed to conduct electricity, and chlorine has no charge.

B Peer assessment

2 Graphite has a melting point of 3642 °C. It is soft and a good conductor of electricity. Explain these three physical properties of graphite in terms of the structure and bonding of graphite. [6]

Student answer

Graphite has a giant covalent bond. It has a very strong bond between metals and non-metals. It takes a large amount of energy to overcome these bonds and make the graphite melt. In graphite there are layers of carbon atoms which are held by only very weak forces which means it is soft. Each carbon is bonded to three other carbon atoms and this means that one electron from each carbon atom is delocalised. Delocalised electrons allow it to conduct electricity.

Use the mark scheme and indicative content to award this answer a level and a mark.

Tip

Before you start marking the answer, approach the question as if you were answering it. Underline the command word and think about what it means, then circle the topic key words.

Mark scheme

Level descriptors	Marks
Level 3 A detailed and coherent explanation is given that demonstrates a good knowledge and understanding of all of the physical properties. It is logically presented.	5–6
Level 2 An explanation is given that demonstrates a reasonable knowledge and understanding of some of the physical properties.	3–4
Level 1 Some simple statements are made. The explanation may be flawed or simplistic.	1–2
No relevant content	0

Indicative content
- High melting point:
  - strong bonds between (carbon) atoms
  - indication that these bonds are covalent
  - substantial energy/heat required to break bonds so melting point is high.
- Soft:
  - layers of carbon atoms that can slide (off/over each other)
  - as there are weak forces of attraction between the layers.
- Conduct electricity:
  - delocalised electrons/free electrons
  - can move and carry the charge.

I would give this a level of .... and a mark of ....

This is because

....

....

....

....

> **Tip**
> ● ● ● ● ● ● ● ● ● ● ● ● ●
> First, assign a level using the level descriptors. Are all three of the physical properties explained? If so, the answer is Level 3.

C Improve the answer

3 The melting points of some substances are shown in the table.

substance	melting point /°C
chlorine (Cl_2)	– 101.5
sodium chloride (NaCl)	801.0

Explain the differences in melting point in terms of the structure and bonding of each substance. [6]

Student answer

In sodium chloride there are sodium ions and chloride ions and these form strong bonds. This makes it hard and rigid. The chlorine does not contain ions. It is a molecule and there are strong covalent bonds in the molecule. The molecule is weak, however, and so it has a low melting point.

Rewrite this answer to improve it and obtain the full six marks.

Extended responses: 'Plan/Design'

The command words 'Plan' or 'Design' mean you simply need to write a method setting out how the experiment will be done.

A Expert commentary

1 Magnesium reacts with dilute hydrochloric acid. Plan an experiment to investigate how the rate of the reaction changes when the concentration of hydrochloric acid is changed. [6]

Student answer

First I would measure out 25 cm^3 using a measuring cylinder of 0.1 mol/dm^3 hydrochloric acid and place it in a conical flask. Quickly start a timer and add magnesium ribbon. Stop the timer and record the time when it is finished. Repeat the experiment using a different concentration of hydrochloric acid, for example 0.2 mol/dm^3. I would keep the temperature the same.

This is a level 2 answer. Award 4 marks.

This is an investigation to compare rates, so the student should state that the mass of magnesium was recorded and the same mass used when repeating.

It is a good idea to mention keeping the temperature constant each time is important as it ensures the investigation is fair.

Good detail is given about the volume and concentration of acid and the apparatus used.

The student should give detail about how they know when the reaction is finished, namely when there will be no more bubbles.

Repeating with one different concentration is not sufficient to determine the effect on rate. Using five different concentrations would be better.

B Peer assessment

2 A student needs to identify a white solid by carrying out some chemical tests. The white solid may be sodium chloride, sodium sulfate, potassium bromide or potassium sulfate. Design an experiment to determine the identity of the solid. [6]

Student answer

Carry out a flame test on all of the four solids and record the results. This should tell you if the solid is a sodium salt or a potassium salt.

You then need to find out if the solids are bromides or sulfates. To do this, you need to dissolve a small sample of each solid in water and then divide into two test tubes. Add some nitric acid and then some barium chloride solution to one test tube and a white precipitate should show the sulfate. Then add some nitric acid to the other test tube. If a cream precipitate forms it is a chloride, while a yellow precipitate indicates a bromide.

Use the mark scheme and indicative content to award this answer a level and a mark.

Mark scheme

Level descriptors	Marks
Level 3 A detailed and logical plan is given that demonstrates a good understanding of the key chemical tests needed to identify the compound.	5–6
Level 2 A plan is given that demonstrates a reasonable knowledge and understanding but lacks some practical detail and does not fully identify the compound.	3–4
Level 1 Some simple statements are made that show a basic knowledge of some chemical tests. The plan may be flawed or simplistic.	1–2

Level descriptors	Marks
No relevant content	0
Indicative content	
• flame test to distinguish between sodium ions and potassium ions	
• yellow/orange flame for sodium ions	
• lilac flame for potassium ions	
• detail of flame test using flame test rod/nichrome wire and Bunsen burner	
• dissolve the substance in water and divide into two samples	
– to one sample add dilute nitric acid and silver nitrate	
– a white precipitate indicates chloride	
– a cream precipitate indicates bromide	
– to the other sample add a few drops of dilute nitric acid and then a few drops of barium chloride solution	
– a white precipitate indicates sulfate ions	

I would give this a level of .... and a mark of ....

This is because

....

....

....

....

C Improve your answer

3 The order of reactivity of the halogens is shown below.

chlorine ↑

bromine | increase in reactivity

iodine |

This order of reactivity can be determined by carrying out displacement reactions using aqueous solutions. Plan an experiment to determine the order of reactivity of these three halogens using displacement reactions. Include details of how the results can be used to determine the order. [6]

Student answer

First, place some potassium iodide solution in a test tube and add some aqueous chlorine. If there is a reaction, then chlorine is more reactive and displaces iodine from solution. In a second test tube you should place some potassium iodide solution and add some aqueous bromine. If there is a reaction, then bromine is more reactive than iodine.

Rewrite this answer to improve it and obtain the full six marks.

Extended responses: 'Evaluate'

In an 'Evaluate' question, you should use the information supplied and your own knowledge and understanding to consider the evidence for and against and draw conclusions.

Your answer is expected to go further than a 'Compare' question, as you need to give a final comment or deduction.

A Expert commentary

1 Diesel is the fuel used by most lorries. Research is being carried out into the use of hydrogen, instead of diesel, as a fuel for lorries. Evaluate the use of hydrogen rather than diesel as a fuel for lorries. [6]

Student answer

This clearly states a reason why hydrogen is good as a fuel.

The raw material to make hydrogen is water and there is an abundant supply of water in the sea. Diesel comes from crude oil. When hydrogen burns it produces water only and so it does not cause any air pollution. Diesel however burns to produce carbon dioxide, which can cause the greenhouse effect. The greenhouse effect causes global warming and ice caps to melt. Incomplete combustion of diesel may produce carbon monoxide which is toxic and also carbon which can cause smog which causes respiratory problems. Hydrogen is produced from water, however as electricity is needed this also may produce pollution.

A better evaluation here would be to continue to state that crude oil is in limited supply and is non-renewable.

More detail about the pollution produced by electricity generation would be useful.

The student has given good detailed evaluation of both fuels in terms of environmental problems caused.

In conclusion, hydrogen is better to use because it is in good supply, and does not cause pollution but it is a flammable gas which is expensive to store safely.

This is good as it draws a conclusion that is consistent with the reasoning in the answer.

This is a Level 3 answer, awarded 5 marks.

Tip

Always try to make a conclusion at the end of an 'evaluate' question.

B Peer assessment

2 The graph shows how the percentage yield of ammonia changes with pressure and temperature.

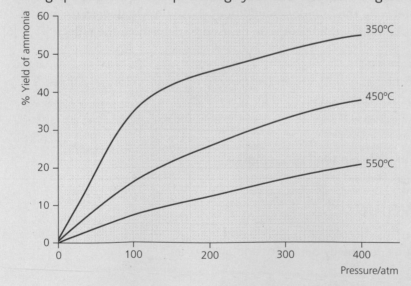

Use the graph and your own knowledge to evaluate the conditions used in industry to manufacture ammonia. [6]

Student answer

A low temperature gives a high yield of ammonia, but a low temperature makes this reaction very slow. Hence a compromise temperature should be used in industry which is moderate and gives a good rate and reasonable yield. Increasing pressure does not have much effect on the yield.

Use the mark scheme and indicative content below to award this answer a level and a mark.

Mark scheme

Level descriptor	Marks
Level 3 A detailed and coherent evaluation is provided that considers different conditions and comes to a conclusion for temperature and pressure consistent with the reasoning.	5–6
Level 2 An attempt to describe some conditions that comes to a conclusion. The logic may be inconsistent at times but builds towards a coherent argument.	3–4
Level 1 Simple statements made. The logic may be unclear and the conclusion, if present, may not be consistent with the reasoning.	1–2
No relevant content	0
Indicative content <ul><li>graph shows that increasing pressure increases yield</li><li>high pressure is expensive due to thick pipes</li><li>compromise pressure of 250 atm is used</li><li>this gives a reasonable yield at reasonable cost</li><li>graph shows that decreasing temperature increases yield</li><li>lower temperature decreases the rate</li><li>in industry a compromise of 450 °C is used</li><li>it is a compromise between a reasonable rate and a reasonable yield</li></ul>	

I would give this a level of .... and a mark of ....

This is because

....

....

....

....

C Improve the answer

3 Evaluate the production of a plastic drink carton with that of a cardboard drink carton in terms of their environmental impact. Use your own knowledge and the information in the table. [6]

	Plastic carton	Cardboard carton
Raw material	crude oil	wood
Temperature used in the process in °C	1600	350
Mass of carbon dioxide produced in kg	0.30	0.50

Tip

To improve this answer, make sure you have mentioned each piece of data in the table. Also try and think of one other piece of information from your own knowledge.

Student answer

The plastic carton is made from crude oil which is a finite resource. The cardboard carton is made from wood. The temperature to produce the plastic carton is twice as high as that to produce cardboard and so more energy is needed. The cardboard carton may be more biodegradable than the plastic carton.

Rewrite this answer to improve it and obtain the full 6 marks.

Extended responses: 'Use'

Some questions may include the command word 'Use' along with another command word. For example, a question may ask you to 'Use' additional information to explain, or to compare. The word 'Use' means your answer must be based on the information or data in the question.

A Expert commentary

1 The table gives some information about the reaction of Group 2 elements with water.

Element	Reactivity with cold water	Name of product
Be	No reaction	No products
Mg	Reacts very slowly with cold water	Magnesium hydroxide and hydrogen
Ca	Reacts moderately with cold water	Calcium hydroxide and hydrogen
Sr	Reacts rapidly with cold water	Strontium hydroxide and hydrogen
Ba	Reacts very rapidly with cold water	Barium hydroxide and hydrogen

Use the information in the table and your own knowledge of Group 1 elements to compare and contrast the reactions of Group 1 and Group 2 elements with water. In your answer, include any trends in reactivity in the groups. [6]

Student answer

A correct trend has been identified using the table.

The student has structured this answer well, writing a point about Group 2 and immediately comparing it to Group 1 in the next sentence.

Another well-structured comparison in two sentences. There is also good use of the data.

The table shows that the Group 2 elements react more vigorously as the group descends. Group 1 elements have the same trend. Group 1 elements include lithium, sodium and potassium and they all react with cold water. However, the table shows that one element in Group 2, beryllium, does not react with water but the other elements in Group 2 do. This is different for Group 1 as all the elements of Group 1 react with cold water.

The table shows that for Group 2, the reaction changes from no reaction to a slow, moderate, rapid, and then very rapid, reaction. This is the same for Group 1. The table shows that the Group 2 metals which react hydrogen gas always produce a hydroxide. This is the same for Group 1 metals.

Using phrases like 'the table shows' indicates that you are using the information given.

This is a Level 3 answer worth 6 marks.

Tip

Read the question carefully. Sometimes you may be asked to use information in the question and also your own knowledge and understanding.

Tip

In a 'Compare' question, you have to write down the similarities and the differences between two (or more) things.

B Peer assessment

2   In an experiment, lithium bromide, sodium bromide and potassium bromide were dissolved in water. The temperature change observed for each solid was measured. The same amount of each compound and water were used each time.

Use the table of data to help you state and explain the results that should be obtained.   [6]

Compund	Energy change when dissolving/kJ
Lithium bromide	−48.8
Sodium bromide	−0.8
Potassium bromide	+19.9

Student answer

For lithium bromide and sodium bromide the energy change is negative. A negative energy change means that the change is exothermic, heat is given out and the temperature has increased in the reaction. For potassium bromide the energy change is positive. This means that the temperature got colder in the reaction as heat is taken in. The results are for three Group 1 metal compounds and they show a trend as you go down the group from lithium to potassium. The trend is that the temperature gets colder. Lithium bromide will show a greater temperature change than sodium bromide. It may also dissolve faster.

Use the mark scheme and indicative content below to award this answer a level and a mark.

Mark scheme

Level descriptor	Marks
Level 3 A detailed and coherent explanation is given that demonstrates a good knowledge and understanding and refers to energy changes and temperature changes for all three substances.	5–6
Level 2 An explanation is given that demonstrates a reasonable knowledge and understanding and refers to energy changes or temperature changes for the three substances.	3–4
Level 1 Some simple statements are made that refer to at least one energy change or temperature change.	1–2
No relevant content	0
Indicative content • lithium bromide and sodium bromide give out energy/heat/are exothermic on dissolving • the temperature should increase when lithium bromide and sodium bromide dissolve • potassium bromide takes in energy/heat and is endothermic when dissolving • the temperature should decrease when potassium bromide dissolves • the energy given out when lithium bromide dissolves is much greater than when sodium bromide dissolves • the temperature change when lithium bromide dissolves is much greater than when sodium bromide dissolves	

I would give this a level of .... and a mark of ....

This is because

....

....

....

....

C Improve your answer

3 The table shows some data about the first four halogens.

Halogen	Melting point /°C	Observation on reaction with hot iron wool
Fluorine	−220	Flames
Chlorine	−101	Glows brightly
Bromine	−7	Glows dull red
Iodine	114	Slowly changes colour

Use your own knowledge and information from the table to state and explain the relative reactivity of the halogens in terms of electronic structure. [6]

Student Answer

From the table I can see that the melting point increases down the group and this may make the elements less reactive. The halogens react more slowly and less vigorously with iron as the group is descended. This means that the reactivity of the halogens decreases down the group. This is because halogen atoms have seven electrons in the outer shell.

Rewrite this answer to improve it and obtain the full 6 marks.

Tip

To improve this answer, first look for any errors then ask yourself if the answer fully explains why the halogens are more reactive down the group.

3 Working scientifically

Working scientifically is the sum of all the activities that scientists do, and it includes several different skills that you need to develop during your study of GCSE Chemistry. These skills fall broadly into four main strands:

1 Development of scientific thinking

2 Experimental skills and strategies

3 Analysis and evaluation

4 Scientific vocabulary, quantities, units, symbols and nomenclature.

This section will just deal with the first three areas, as the fourth area – vocabulary, quantities, units, symbols and nomenclature – is covered in the Maths section of the book.

Your GCSE paper will include questions that assess all these strands and you should expect to see questions that:

● require knowledge of how scientists have found out what they know
● directly assess practical and enquiry skills.

» Selecting apparatus and techniques

As part of your GCSE Chemistry course, you are required to demonstrate your capability in using a range of apparatus and techniques (AT skills). These are summarised in Table 3.1. Exam questions may test your knowledge of how these AT skills can be applied to your studies.

Table 3.1 GCSE Chemistry AT skills

List of apparatus and techniques	
AT1	*Use of appropriate apparatus to make and record a range of measurements accurately, including mass, time, temperature and volume of liquids and gases. See pages 65–66 for more information on this skill.*
AT2	*Safe use of appropriate heating devices and techniques, including use of a Bunsen burner and a water bath or electric heater.*
	To show that you can safely use heating devices you should be aware that some chemicals are flammable and so a flameless method is used to heat them. Instead of using a Bunsen burner directly, a hot water bath or an electric heating mantle are used.
	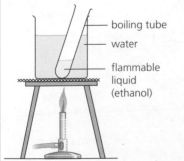 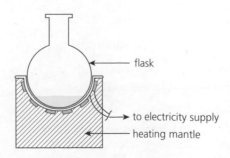 ▲ Figure 3.1 Flameless methods of heating

AT3	*Use of appropriate apparatus and techniques for conducting and monitoring chemical reactions, including appropriate reagents and/or techniques for the measurement of pH in different situations.*
AT4	*Safe use and careful handling of gases, liquids and solids, including careful mixing of reagents under controlled conditions, using appropriate apparatus to explore chemical changes and/or products.*

For **AT3** and **4**, you will need to know the exact procedure for different experiments such as titrations, rates experiments, preparation and collection of gases and measuring pH using a pH probe or universal indicator paper.

AT5	*Safe use of a range of equipment to purify and/or separate chemical mixtures, including evaporation, filtration, crystallisation, chromatography and distillation.*

In addition to using different apparatus and techniques you should make sure that you can draw scientific diagrams (where appropriate) to show the apparatus set up for each of the following techniques shown.

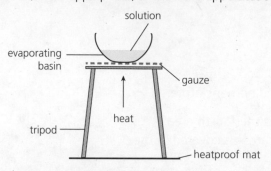

▲ **Figure 3.2 Evaporation**

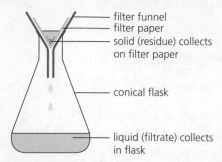

▲ **Figure 3.3 Filtration**

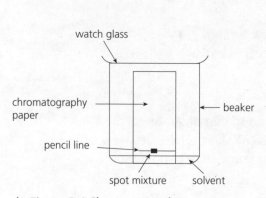

▲ **Figure 3.4 Chromatography**

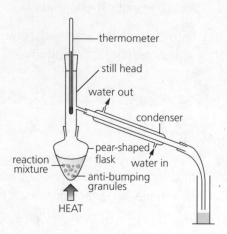

▲ **Figure 3.5 Distillation**

Crystallisation occurs when a saturated solution in an evaporating basin is cooled and so the solute crystallises out of the solution as crystals.

AT6	*Making and recording of appropriate observations during chemical reactions, including changes in temperature and the measurement of rates of reaction by a variety of methods such as production of gas and colour change.*

For this skill, you need to be able to use and draw apparatus set up as shown in Figures 3.6, 3.7 and 3.8, and record measurements to help you determine rate of reaction.

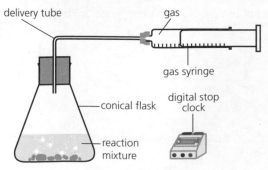

▲ **Figure 3.6** Experimental set up to measure reaction rate by measuring volume of gas as time proceeds

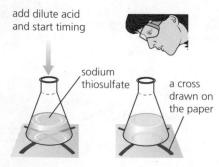

▲ **Figure 3.7** Experimental set up to measure reaction rate by changing concentration through observation

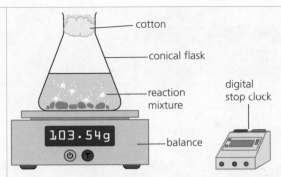

▲ Figure 3.8 Experimental set up to measure reaction rate by mass decrease

AT7	*Use of appropriate apparatus and techniques to draw, set up and use electrochemical cells for separation and production of elements and compounds.*

For the topic 'Cells and electrolysis', you need to recognise the symbols for a battery and a lamp, as shown in Figure 3.9.

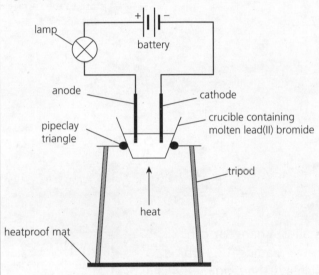

▲ Figure 3.9

AT8	*Use of appropriate qualitative reagents and techniques to analyse and identify unknown samples or products, including gas tests, flame tests, precipitation reactions, and the determination of concentrations of strong acids and strong alkalis.*

For this skill, you will need to be able to:
- describe how to test for gases such as hydrogen, carbon dioxide and oxygen,
- use a flame test to identify metal ions, e.g. potassium ions have a lilac flame test,
- describe adding nitric acid and silver nitrate to test for halide ions or nitric acid and barium chloride to test for sulfate ions by forming a precipitate,
- describe how to carry out a titration between an alkali and acid.

AT 1, recording measurements accurately by using appropriate apparatus

For this skill you need to be able to choose the correct piece of apparatus to measure different chemicals. Some pieces of apparatus that you need to be familiar with are given in Table 3.2.

Table 3.2 Examples of measurements and instruments used

Measurement	Measuring instrument used
Volume of liquids	Measuring cylinder (accurate to 1 cm³) Pipette with safety pipette filler (accurate to 0.1 cm³) Burette (accurate to 0.05 cm³)
Volume of gases	Measuring cylinder or burette (under water) (cm³) Gas syringe (cm³)
Mass	Balance (g)
Temperature	Thermometer (°C) Temperature probe (°C)
Time	Digital stopwatch (s)

When using a measuring cylinder, a meniscus is the curve seen at the top of a liquid in response to its container. When reading the volume of the liquid, the measurement at the bottom of the meniscus curve is read at eye level.

Key term

Meniscus: The curve that is seen at the top of a liquid close to the surface of the container.

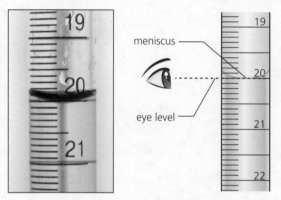

meniscus

eye level

▲ Figure 3.10 Using a measuring cylinder

The resolution of a piece of apparatus is the smallest change it can measure. For example, the resolution of the burette reading 12.3 cm³ is ±0.1 cm³ and the resolution of a balance that displays the reading 8.45 g is 1×10^{-2} (0.01) g.

When a measurement is made there is always some doubt or uncertainty about its value. Uncertainty is often recorded after a measurement as a ±. The uncertainty can be estimated from the range of results that are obtained when an experiment is repeated several times.

Key terms

Resolution: The smallest change a piece of apparatus can measure.

Uncertainty: The range of measurements within which the true value can be expected to be.

(A) Worked example

1   The volume of a gas produced in a reaction was measured five times. The results were 82, 77, 78, 96 and 80 cm³. Find the mean volume.

Model answer

The mean value is found after excluding any anomalous results. 96 cm³ is anomalous as it is significantly different from all the others.

$$\text{Mean value} = \frac{82 + 77 + 78 + 80}{4} = 79 \pm 3\,\text{cm}^3$$

The mean is quoted to the nearest unit as all the values are measured to the nearest unit. The uncertainty is ±3 cm³ as the highest and lowest values are within 3 cm³ of the mean.

Tip

Uncertainty is the difference between the mean value and the value furthest from the mean. For more information on uncertainty, see page 75.

» The development of scientific thinking

This area of working scientifically is about understanding how scientific theories come to be developed and refined as well as recognising the importance of working safely and the limitations of what we can discover.

Development of scientific methods, theories and models

Every day we observe the world around us and use our senses to detect what is happening. As scientists, we use observations to ask questions about why things happen. For example, why does water evaporate faster than oil, or which tissue paper is more absorbent and why? Observations can then lead to a hypothesis, which is really an idea about how or why something happens, such as, 'water evaporates faster than oil as it is less viscous'. Scientists carry out experiments to test hypotheses or models and draw conclusions and report on their work.

Models are used in science to help explain how something works or to describe how something is structured. They can also be used to make predictions or to explain observations. However, a model is never exactly like the real thing (if it were, it would no longer be a model). There are many kinds of scientific models. Some examples are physical models – such as ball and stick models, mathematical models, and conceptual models. Models do have limitations but are useful and can be changed based on new evidence.

The whole process that scientists use to show whether their ideas or models are correct or not is shown in Figure 3.11.

If a hypothesis is proved to be correct, then it becomes a theory. Scientists build on these theories by asking more questions and so continuing the process; this is how theories develop over time and advances in science are made. Using the scientific method, breakthroughs are made in medicine, drug discovery, forensic, environmental chemistry and material science.

The Periodic Table illustrates how scientific theories change over time and how the work of each scientist builds on that of the previous one. For example, Döbereiner's triads led to Newlands' octaves and then to Mendeleev's layout, which we still use today.

> **Key term**
> Model: This is a representation of a thing or process in a way that aids understanding.

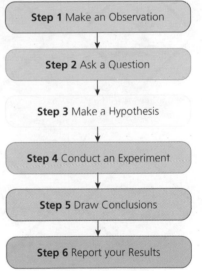

▲ Figure 3.11 The scientific method

Appreciating the limitations of science and ethical issues

Any new development in science is exciting and can be used to advance technologies and introduce new methods and chemicals. However, scientific developments can have drawbacks as well as benefits, and even risks to the environment and mankind. For example, the effect of some nanoparticles on the human body has not been fully discovered and in the future scientists may find that they can cause risks to our health. Some new developments may also cause ethical problems; for example, some people feel it is wrong to experiment on human embryos.

> **Key terms**
> Hypothesis: A proposal intended to explain certain facts or observations.
>
> Theory: This is a hypothesis that has been proved by experiment or observation to be correct.

Before a new scientific development is used, the factors listed here must be considered:

- cost
- the effect on the environment
- the effect on people
- ethics.

You can see that not all the factors used to make a decision are based on science and there are limitations on how much science can be used. These issues are decided on by individuals, governments and society.

Evaluating the risks in practical experiments

Whenever experiments and investigations are carried out in the laboratory you need to decide if the experiment is safe by carrying out a risk assessment.

A risk assessment should include:

- a list of all the hazards in the experiment
- a list of all the risks that the hazards could cause
- suitable safety precautions you should take to reduce or prevent the risk.

Chemicals should have a COSHH hazard warning sign on the container. The ones you should recognise are shown in Figure 3.12.

Dangerous to the environment	Toxic	Gas under pressure
Corrosive	Explosive	Flammable
Caution – used for less serious health hazards like skin irritation	Oxidising	Longer-term health hazards such as carcinogenicity

▲ Figure 3.12 Hazard warning symbols

Key terms

Ethics: This is the consideration of the moral right or wrong of an action.

Risk assessment: A judgement on how likely it is that someone might come to harm if a planned action is carried out and how these risks could be reduced.

Tip

You will need to do a risk assessment when completing the required practicals for your GCSE Science course.

Some examples of hazards, risks and safety precautions are shown in Table 3.4.

Table 3.4 Hazards, risks and safety precautions

Hazard	Risk	Safety precaution
Concentrated acid	Corrosive to eyes and skin	• Wear safety glasses • Wear gloves • Use small amounts
Ethanol	Flammable	• Keep away from Bunsen and flames • Use a water bath or electric heater to heat
Bromine	Toxic	• Wear gloves • Wear safety glasses • Use small amounts • Use fume cupboard
Potassium/sodium	Explosive	• Use small amounts • Store under oil • Wear safety glasses and use a screen
Cracked glassware	Could cause cuts	• Check for cracks before use
Hot apparatus	Could cause burns	• Allow to cool before touching • Use tongs
Heating chemicals in test tubes	Chemical could spit out	• Wear safety glasses • Point test tube away from others
Long hair	Could catch fire	• Tie back long hair

> **Tip**
> • • • • • • • • • • • • •
> Wearing safety glasses is a requirement for *all* Chemistry practicals.

Recognising the importance of peer reviews

There are many things that a scientist must do before valid results and conclusions from experimental results can be made:

- the experiment must be repeated and similar data obtained.
- the experiment must be repeated under different conditions, or by using a slightly different method so that the results are reproducible.
- before publication of results and conclusions, scientists must present their ideas at conferences and have their work checked and evaluated by other experts in the same field. This process is known as peer review. Scientific findings are only accepted once they have been evaluated critically by other scientists.

> **Key terms**
>
> Valid results: Results or data obtained from an appropriately designed experiment.
>
> Reproducible: This is a result that can be repeated by another person or using a different technique.
>
> Peer review: The evaluation of scientific work by others working in the same field.

Questions

1. A student wishes to use a Bunsen burner to heat a beaker of ethanol. Explain one suitable precaution, other than wearing eye protection, to reduce the risk of harm in this procedure.

2. A hypothesis for an experiment states that the solubility of potassium nitrate depends on the temperature of the water.
   a. Write down two general scientific method points that should be carried out after the hypothesis is written.
   b. Potassium nitrate is an irritant. Complete the risk assessment for the use of potassium nitrate in this practical.

Risk	Control measure
Potassium nitrate powder is an irritant	

3. a. In a practical, a piece of potassium is placed in a trough of water and the observations recorded. Write down three things that could be done to control the risks in this experiment.
   b. The practical was repeated using a piece of sodium of the same mass to determine which metal reacted faster with water. Write a hypothesis, based on your knowledge of the reactivity series for this experiment.

4 In some areas, fluoride is added to the tap water supply.
   a State one advantage of adding fluoride to tap water.
   b Suggest one reason why people may object to adding fluoride to water.
5 The results of drug trials are peer reviewed before they are published. Why are peer reviews important in drug trials? Select one answer.
   A To calculate the best dose needed.
   B To check the drug works.
   C To make sure that the results are evaluated and correct.
   D To make sure the scientist gets credit for making the drug.
6 To prepare a sample of an ester 5 cm³ of ethanol and 5 cm³ of ethanoic acid were added to a test tube with 5 drops of concentrated sulfuric acid catalyst. The mixture was heated for a few minutes. Write a risk assessment for this practical.
7 Scientific research has shown a correlation between silver nanoparticles and the speed of healing of a cut. Explain how the scientific community would validate this research.

» Experimental skills and strategies

To be a good scientist you need to develop the experimental skills and strategies that will be explored in the following sub-sections.

Writing hypotheses

Most scientific investigations begin with an observation, for example, you may observe that a lump of sugar dissolves more slowly than granules of sugar. The first step in the scientific method is to write a hypothesis for an observation. A hypothesis is really a suggestion about how or why something happens.

The phrase 'depends on' is often used when writing a hypothesis, for example 'the speed of dissolving *depends on* the surface area of the sugar'. From this you can predict that a lump of sugar will dissolve faster/slower than the same mass of sugar granules. Scientific knowledge can be used to back up a hypothesis and an experiment can be planned to test it.

Planning hypotheses

To test a hypothesis, an experiment must be planned using different variables. Variables are the factors that can be changed during an experiment. They may be, for example, temperature, mass, volume, pH or even the type of chemical used in an experiment. When planning experiments, you should only change one variable while keeping others the same.

> **Key term**
> Variable: A characteristic or chemical quantity.

There are different types of variables, which you need to be familiar with.

Table 3.5 Types of variable

Type of variable	Description
Continuous variable	It has values that are numbers.
	Mass, temperature and volume are examples of continuous variables.
	The values of these variables can either be found by counting (for example, the number of drops) or by measurement (for example, the temperature).
Categoric variable	It is best described by words.
	Variables such as the type of acid or the type of metal are categoric variables.

> **Key terms**
> Continuous variable: A variable that has a numerical value.
>
> Categoric variable: A variable that has a value that is a label, e.g. the name of an acid.

Type of variable	Description
Scientists often plan experiments to investigate if there is a relationship between two variables, the independent and the dependent variable.	
Independent variable	The variable for which values are changed or selected by the investigator; i.e. it is the one that you deliberately change during an experiment.
Dependent variable	The variable that may change as a result of changing the independent variable. This is the variable that is measured for each and every change in the independent variable.
Control (or controlled) variable	The controlled variable may, in addition to the independent variable, affect the outcome of the investigation. Control variables must be kept constant during an experiment to make it a fair test.

Key terms

Independent variable: The variable that is deliberately changed in an experiment.

Dependent variable: The variable that is measured whenever there is a change in the independent variable.

Control (or controlled) variable: The variable(s) that is not either the dependent or independent, and that must be kept constant during an experiment.

Once the variables for an experiment have been determined, an experimental method needs to be written using a range of apparatus and techniques.

For a practical investigation to be a fair test that produces valid results, only two variables should change – the independent variable (the one which we change) and the dependent variable (the one which changes because we have changed the independent variable). Everything else should remain the same, i.e. every other variable should be controlled. In other words, a fair test is one in which only the independent variable has been allowed to affect the dependent variable.

Recording observations and measurements

During experimental activities you will often record results in a table. When drawing tables and recording data ensure that:

- the table is a ruled box with ruled columns and rows.
- there are headings for each column and row.
- there are units for each column and row – usually placed after the heading after a solidus (/) or in parentheses ( ), for example, 'Temperature /°C' or 'mass (g)'. Units should not be written in the body of the table.
- there is room for repeat measurements and averages – remember, the more repeats you do, the more reliable the data.
- the independent variable is recorded in the first column and the dependent variable(s) can be recorded in the next column(s).
- data should be recorded to the same number of decimal places or significant figures.

Qualitative observations are what we see and smell during reactions. Important types of observations in Chemistry and notes on how to record these are shown in Table 3.6.

Tip

Remember that clear is not a colour – instead use the word 'colourless'. For example, hydrochloric acid is a *colourless* liquid. It is also clear, but this refers to the fact that it is transparent.

Table 3.6 Types of observation

Type of observation	Notes on recording observations	Examples
Colour change	Always state the colour of the solution before the reaction and after.	*When bubbling an alkene into bromine water – the colour change is an orange solution to a colourless solution.*
Bubbles produced	If a gas is produced, then bubbles are often observed in the liquid and the solid reactant disappears.	*When sodium carbonate reacts with an acid, the observation is bubbles and the solid reactant disappears* (Note, writing that 'carbon dioxide is formed' is **not** an observation.)

Type of observation	Notes on recording observations	Examples
Heat produced	In many reactions the temperature changes.	*When acids react with alkalis, the temperature increases.*
Precipitate produced	When two solutions mix, often an insoluble precipitate forms. Ensure you use the word 'precipitate' in your observation, as a common mistake is to write that the solution becomes cloudy. Also, do state the colour of the precipitate and the colour of the solution before adding the reagent.	*When barium chloride solution is added to a solution containing sulfate ions, a white precipitate is formed in the colourless solution.*
Solubility of solids	When a spatula of a soluble solid is added to water, the observation is often that the solid dissolves to form a solution. Make sure you state the colour of the solution formed.	*Copper(II) sulfate crystals dissolve in water to produce a blue solution.*
Solubility of liquids	When a liquid is added to water, always record whether it is miscible or immiscible with water.	*Ethanol and water are miscible.*

Evaluating experimental methods

In everyday life, we are constantly evaluating situations and drawing conclusions. For example, when baking muffins, we might evaluate how the finished muffins look and decide how to improve on our method. If the muffins are

- too pale, then they probably should have been left in the oven for a longer time
- burnt around the edges, then a cooler oven should have been used or perhaps the muffins should have been removed from the heat sooner
- not moist enough, then perhaps some extra liquid ingredients should have been added or maybe less flour should have been used.

Evaluating an experimental method allows you to assess its effectiveness, plan for future modifications and judge whether an alternative method might be more suitable. Part of the evaluation procedure includes asking questions such as:

- Was your method suitable?
- Were there any sources of error? How could you improve your experiment to eliminate sources of error?

For example, if an experiment is to measure the temperature change during a neutralisation reaction, a source of error may be that heat is lost from the sides of the plastic cup or from the open top. The method could be improved on by using a lid on the cup and insulating the side with cotton wool.

> **Key term**
>
> Evaluate: This means to weigh up the good points and the bad points.

> **Key term**
>
> Error: The difference between an observed value and what is true in nature. Error causes results that are inaccurate or misleading.

Questions

1   For the following investigations, identify the:
    - independent variable
    - dependent variable
    - controlled variables.
    a   Some magnesium was added to hydrochloric acid and the temperature recorded. The experiment was repeated several times using different volumes of hydrochloric acid.   [3]
    b   In the reaction between copper carbonate and hydrochloric acid, the time taken for a mass of copper carbonate to be completely used up was recorded. The experiment was repeated using different masses of copper carbonate.   [3]
    c   The temperature of nitric acid was recorded before and after some sodium hydroxide was added. The experiment was repeated using sulfuric acid, ethanoic acid and methanoic acid.   [3]
2   In an investigation, 2 g of copper(II) sulfate crystals were found to dissolve faster in hot water than in cold water.
    a   Write a hypothesis for this investigation.
    b   State an observation that occurred in the investigation.
    c   Name three pieces of apparatus that would be used in this investigation.

3 Copy and complete the diagram to show how you can distil copper(II) sulfate solution and collect pure water. Label the pure water and the copper(II) sulfate solution.

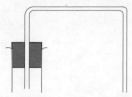

4 In an experiment, a student weighed some hydrated magnesium sulfate crystals, heated them for 2 minutes and reweighed them. The results recorded are shown in the table.

Mass of crucible and hydrated magnesium sulfate before heating /g	Mass of crucible and magnesium sulfate after heating
9.37	8.25

  a   What is the resolution of the balance used by the student?

  b   Suggest one improvement that could be made to the results table.

5 In a laboratory experiment a student added some magnesium metal to some copper(II) sulfate solution and recorded the observation that 'the magnesium became covered in copper and magnesium sulfate was formed'.

  a   Write a word equation for the reaction occurring.

  b   State and explain if the observations given by the student are correct.

6 The diagram shows apparatus used to investigate the reaction of hydrochloric acid and calcium carbonate.

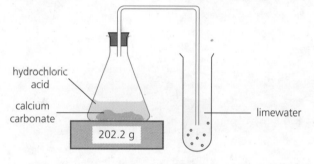

  a   Describe and explain the observations that would occur.

  b   What is the resolution of the balance?

7 50 cm³ of hydrogen peroxide and 1.0 g of manganese dioxide was allowed to react at 25 °C. The volume of oxygen collected from the reaction at 10 second intervals is: after 10 seconds, 30 cm³ after 20 seconds, 49 cm³ after 30 seconds, 59 cm³ after 40 seconds and 63 cm³ after 50 seconds. On repeating the experiment, the volume of gas obtained at each time interval was 32, 51, 59, 63, 65 cm³, respectively.
Present these results in a suitable table with headings and units. Calculate and record the average volume of gas produced.

» Analysis and evaluation

Analysing data

When carrying out experiments, you often record your results in a table. You must also be able to translate data from one form to another. In Chemistry, this often involves using data from a table to draw a graph. A graph is an illustration of how two variables relate to one another. It is often easier to see patterns and trends by using a graph to show results of an experiment. For more information on drawing graphs, check back to page 33 of the maths chapter.

Interpreting and explaining data

A graph can help determine if there are any changes in the dependent variable as the independent variable changes. Instead of using words to tell us what is going on, a graph uses shape. To describe what a graph is showing you really need to 'Tell the story of the line' in terms of the variables. For example, the graph in Figure 3.13 could have a description such as 'the line is horizontal showing that the mass of catalyst does not change as time increases'.

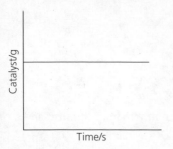

▲ Figure 3.13

A graph may be in sections, and you should be able to describe what each section is showing. Always start reading the line graph on the left-hand side – just like reading a sentence.

To interpret a graph:

- state if there is a general trend across the graph – do the variables increase or decrease in relation to each other?
- describe the trend in each section of the graph,
- use data from the graph in your answer.

The graph in Figure 3.14 has three different sections.

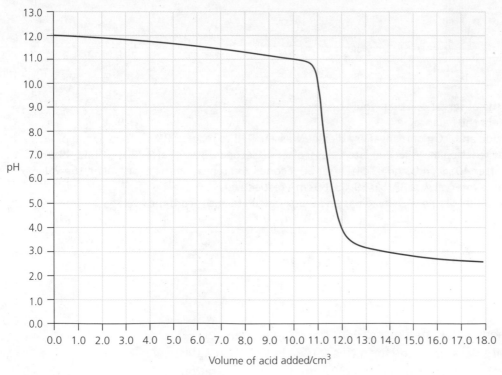

▲ Figure 3.14

A good description of this graph is:

Section 1: The pH falls very gradually from 12.0 to 11.0 as the volume of acid added is increased from $0.0\,cm^3$ to $10.8\,cm^3$.

Section 2: The pH then falls steeply from 11.0 to 3.5 as the volume of acid is increased to a total of $12.0\,cm^3$.

Section 3: The pH then falls gently from 3.5 to 2.7 as the volume of acid increases from $12.0\,cm^3$ to $18.0\,cm^3$.

Evaluating data

When evaluating data obtained from an investigation, you should try and identify errors. Errors may be random or systematic.

A random error is one that causes a measurement to differ from the true value by different amounts each time. Random errors are often the result of a poor measurement being made, usually due to the method not being carried out consistently. The results would vary in an unpredictable way from one measurement to the next. The effect of random errors can be reduced by making more measurements and calculating a new mean.

A systematic error is one which causes a measurement to differ from the true value by similar amounts each time. Systematic errors can occur when repeated parts of an investigation are done in the same way, but that way is not correct, that is, the same problem is affecting the result each time. Sources of systematic errors can include the environment, methods of observation or the instruments used. Systematic errors cannot be corrected by simply repeating the experiment in the exact same manner. If a systematic error is suspected, the experiment should be repeated using a different technique or a different set of equipment, and the results compared.

You also need to evaluate the results by asking the following questions:

1 Are the results accurate?
2 Are the results reproducible?
3 Are the results repeatable?

Results are only accurate if they are close to the true value. To get accurate results. You can:

- repeat your results
- repeat using a different measuring instrument to determine whether you get the same readings
- use high-quality instruments that measure accurately, e.g. a thermometer that measure to one decimal place
- ask yourself whether the results are precise.

Results are precise if all the results are close to the mean value, that is, there is little spread around the mean value. The mean of the results shown in Table 3.7 is $15.0\,cm^3$. The results are precise as they are all close to the mean. See pages 22–23 for information on calculating mean values.

Table 3.7 Experimental results

Volume of gas produced /cm³	Experiment 1	Experiment 2	Experiment 3
	14.8	15.3	14.9

Results are reproducible if similar results are obtained when an experiment is repeated by another person or by using a different technique.

Results are repeatable if similar results are obtained when an experiment is repeated by the same person several times.

Reporting your findings

Scientists must be able to communicate the findings of their investigation clearly and correctly. Many scientists do this by writing scientific papers. A paper will include the nature of the investigation, the method and the data obtained, which

> **Key term**
>
> Random error: This is an error that causes a measurement to differ from the true value by different amounts each time.

> **Key term**
>
> Systematic error: An error that causes a measurement to differ from the true value by the similar amounts each time.

> **Key terms**
>
> Accurate: Something that is close to the true value and repeatable.
>
> Precise: These are results that are close to the mean value where there is little spread around a mean value.

> **Tip**
>
> Remember that uncertainty is the difference between the mean value and the value furthest from the mean. The uncertainty of the results in the table is $15.3 - 15.0 = \pm0.3\,cm^3$

> **Key term**
>
> Repeatable: These are results that can be obtained by the same person repeating an experiment.

may be presented in tables and in graphical form. It will also interpret the results and evaluate them. Scientific papers are presented to other scientists and then peer assessed. If the paper is verified, then it is published in a scientific journal.

Questions

1 Two students carried out experiments to find the percentage by mass of nitrogen in ammonium sulfate. The true value is 21.2%.

| Student A | 21.4% | 21.2% | 21.1% | 21.3% | 21.4% |
| Student B | 22.5% | 22.4% | 22.6% | 22.5% | 22.5% |

   a Calculate the mean value for each student. State the uncertainty in the mean.
   b Comment on the accuracy of the mean result for each student.
   c Comment on the repeatability of the result for each student.
   d Why did the results differ when each student repeated them?
   e Explain whether either of the students had a systematic error in their experiment.

2 In an experiment, some calcium carbonate and acid were placed in a conical flask on a balance and the balance reading was recorded every minute. The results were recorded and the graph shown below was drawn.

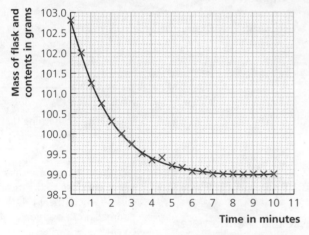

   a Are there any results that you would ignore when drawing a best-fit curve?
   b At a time of 2 minutes, what is the mass of the flask and contents?
   c Describe the trend in the graph.
   d How can the effect of random error be reduced in this practical?
   e How would the effects of systematic errors be reduced in this practical?

3 A student carried out experiments and made the errors listed next. Decide whether these errors lead to random error or systematic error in the results.
   a Using a balance that has a piece of plastic tape stuck to the pan on the top.
   b Forgetting to re-zero the balance on one repeat of the experiment.
   c Using a burette in a titration that has a slight leak at the tap.

4 Revision skills

This section covers the importance of revision and the key strategies that can help you gain the most benefit from your revision. A common misconception is that there is only one way to revise – one that involves lots of note-taking, re-reading and highlighting. However, research shows that this is not an effective way of revising. You need to vary the techniques you use – and find the ones that work best for *you* – to make the most of revision.

Students often think they can change the way they revise, or that revision is something you either can or cannot do. In fact, revision is an important skill and, like any skill, with support and practice you can get better at it. By 'better', this means that you can revise more efficiently (in other words, you'll get a greater benefit from the same amount of revision time) and more effectively (in other words, you'll retain more information).

This chapter will cover the key elements of successful revision:

- Planning ahead
- Using the right tools
- Creating the right environment
- Useful revision techniques
- Practice, practice, practice!

» Planning ahead

The key to successful revision is planning. There are a number of things to bear in mind when planning revision.

Be realistic

There is nothing more demotivating than setting unrealistic targets and then not fulfilling them. You need to think carefully about how much work you can realistically complete and set a reasonable time to complete it.

Ensure you cover all topics in the course

It is tempting to focus on what you think are the most important areas and leave out others. This is risky because no one knows what you will be asked about. It's a horrible feeling seeing an exam question on a topic that you know you haven't revised. In this section, there is advice on the sorts of strategies to ensure you cover all the key points of the specification.

Make friends with the areas you don't like

It is tempting to focus on the areas you already know and are good at. It makes you feel like you're making great progress when, in fact, you're doing yourself a disservice. You should work hard at the areas you find difficult to make sure you give yourself the best chance. This can be tough as you may feel progress is slow, but you must persevere with it.

> **Tip**
> Spend a small amount of time each evening during your GCSE course going over what you learnt in that day's lesson – it can be really beneficial. It helps you remember the content when you come to revise it, and provides good preparation for the next lesson.

Ask for help

The most successful students are often those who ask questions from teachers, parents and other students. If there is anything on the specification that you are unsure about, don't stay silent – ask a question! Proper planning will ensure you have time to ask these questions as you work through your revision.

Target setting

Targets are an important part of successful revision planning. You may want to include SMART targets in your revision timetable.

Here's an example of a SMART (specific, measurable, achievable, realistic and timely) target.

Target: Achieve at least a grade 6 in a practice Chemistry Paper 1 done under exam conditions. This should be completed by the end of the week.

- **Specific** – this target is specific as it gives the exam paper, how it needs to be completed and the grade required.
- **Measurable** – as a specific minimum grade is given (6), this target is measurable.
- **Achievable** – as long as there is time to complete the paper, which there should be if it's being completed in the 'time allowed', then this target would be achievable.
- **Realistic** – you shouldn't be expecting to score grade 9 in assessments straight away or learn huge amounts of content in a very small time; so, a grade 6 seems to be realistic for a first stab.
- **Timely** – there is a set time to complete this goal, namely by the end of the week. Assuming that the student has revised all of the topics on this paper by then, this is a sensible timeframe.

Targets can also be smaller and set for individual revision sessions, for example:

- complete three practice questions on one maths skill
- get 75% on a recall test
- learn the stages of a process, e.g. the carbon cycle
- make a set of key word flash cards on Lenses and Visible Light

Setting targets for each revision session will help you realise when you are finished, as well as providing yourself with evidence of your progress – always a good motivator!

> **Tip**
> Targets can include things such as not using social media or your phone for a whole revision session if this is something you particularly struggle with.

» Using the right tools

Having the right tools is vital for effective revision. Some of the 'practical' tools you'll need during your revision would include:

- a planner or diary
- pens
- paper
- highlighters
- flash cards
- and so on ...

Having these tools close to hand will remove simple barriers to successful revision – such as not having a pen!

Revision timetables

Revision timetables are a useful tool to help you organise and structure your work. Remember that the key is to be realistic – don't plan to do too much, or you'll become demoralised.

Revision works best in shorter blocks. So, don't plan to spend two hours solidly revising one topic – you probably won't last that long. Even if you do, it's unlikely the work towards the end of this time will be effective.

If you are making a revision timetable for mock exams (before you've finished your course), you will need to allow time for any homework set in addition to revision.

How to create a revision timetable

Identify the long-term goal and short-term targets you're trying to achieve (and make sure they're SMART). Ask yourself if this a general timetable to use during the term, or one aimed at preparing for a particular exam or assessment. This will affect how you build your plan as your commitments will vary.

Whatever the end goal, don't plan so you only just finish in time. Make sure you plan to cover all the topic areas you need well before the assessment. That way, if you encounter problems that slow you down, you won't run out of time.

Examples of revision timetables

Good example

Revision sessions split into small sections. This helps maintain engagement during the session.

Regular breaks scheduled and realistic expectations of how much revision can be completed in a day.

Times	Mon
8:30am–3:20pm	School
4:00pm–4:30pm	Chemistry (size and mass of atoms)
4:30pm–5:30pm	Football
5:30pm–6:00pm	Dinner
6:00pm–6:30pm	Physics (black body radiation)
6:30pm–7:00pm	Online gaming
7:00pm–7:30pm	Biology (meiosis)

Bad example

Unrealistic expectations – timetabling so revision starts at 6:30 am and finishes at 11:00 pm at night is unrealistic and potentially harmful. Failing to achieve set goals can be very demotivating.

Working excessive long hours without adequate sleep and relaxation time can be detrimental to health.

Times	Mon
6:30am–7:20am	Physics
8:30am–3:30pm	School
3:30pm–5:00pm	Physics
5:00pm–7:30pm	Chemistry
7:30pm–11:00pm	Biology

No breaks scheduled – planning breaks, both as a rest and reward, are very important for effective revision.

> **Tip**
> Include your other commitments in a revision timetable, such as music lessons, sports, exercise or part-time work. This will give a clearer picture of how much time you have for revision. These commitments could be rewards – they give you something look forward to. Or it may become clear that you may have too much on and have to (temporarily) give something up.

> **Tip**
> Make sure you carefully plan how much time you have available before each exam. Miscalculating by even a week could cause problems.

Specific topics given for revision sections – while you don't need to necessarily rigidly stick to this it's good to have a topic focus for each revision session, you can then set targets for the session based around this particular topic area.

No specific topics mentioned – 'Physics' is far too vague; what areas are they specifically going to work on?

Long blocks of one subject – the student is unlikely to remain engaged for this length of time.

Revision checklist

A revision checklist is an important tool to ensure you are covering all the required specification content. Your teacher may provide you with a revision checklist, but even if they do, making one yourself can be a useful learning activity.

Tip

Some revision guides (like *My Revision Notes*) also have checklists included that you can use.

How to make a revision checklist

1 Read the specification; this is everything you need to know.
2 Split the specification into short statements and place them into a grid.
3 Work through the grid, ticking as you complete each stage for a particular topic. Use practice exam questions to check that your revision has been effective.
4 Return to the areas you are weaker in and focus on improving them.

Example revision checklist

The following is an example statement taken from a GCSE Physics specification. This statement has been used as the basis for an example revision checklist.

Learners should have a knowledge and awareness of the advantages and disadvantages of renewable energy technologies (e.g. hydroelectric, wind power, wave power, tidal power, waste, solar, wood) for generating electricity. Learners should also be able to explain the advantages and disadvantages of non-renewable energy technologies, including fossil fuels and nuclear for generating electricity.

Revision checklist

Specification statement	Covered in class	Revised	Completed example questions	Questions to ask teacher
Advantages and disadvantages of renewable energy resources for generating electricity 1 – hydroelectric, wind power, wave power, tidal power.				
Advantages and disadvantages of renewable energy resources for generating electricity 2 – waste, solar, wood.				
Advantages and disadvantages of fossil fuels for generating electricity.				
Advantages and disadvantages of nuclear power for generating electricity.				

Posters

You could create posters of key processes, diagrams and points and put them up around the house so you can revise throughout the day. Be sure to change the posters regularly so that you don't become too used to them and they lose their impact. See the next section for more on making the most of your learning environment.

Technology

There are many ways to use technology to help you revise. For example, you can make slideshows of key points, watch short videos or listen to podcasts. The advantage of creating a resource yourself is that it forces you to think about a particular topic in detail. This will help you to remember key points and improve your understanding. The finished products should be kept safe so you can revisit them closer to the exam. You could lend your products to friends and borrow ones they've made to share the workload.

Tip

Don't procrastinate by focusing too much on the appearance of your notes. It can be tempting to spend large amounts of time making revision timetables and notes that look nice, but this is a distraction from the real work of revising.

Making your own video and audio

If you record yourself explaining a particular concept or idea, either as a video or podcast, you can listen to it whenever you want. For example, while travelling to or from school. But make sure your explanation is correct, or you may reinforce incorrect information.

Revision slideshows

Slideshows can incorporate diagrams, videos and animations from the internet to aid your understanding of complex processes. They can be converted into video files, printed out as posters, or viewed on screen. It's important to focus on the content of the slideshow – don't spend too long making it look nice.

Social media

Social media contain a wide range of revision resources. However, it is important to make sure resources are correct. If it's user-generated content, there's no guarantee the information will be accurate.

Study vloggers and other students on social media can provide valuable support and a sense of being part of a wider community going through the same pressures as you. However, don't compare yourself to other people in case it makes you feel as if you're not keeping up.

» Creating the right environment

The importance of having a suitable environment to revise in cannot be underestimated – you can have the best plan and intentions in the world, but if you're watching TV at the same time, or you can't find the book you need, or you're gasping for a drink and so on, then you're likely to lose concentration sooner rather than later. Make sure you create a sensible working space.

Work area and organisation

It is difficult to concentrate with the distraction of an untidy work area – so keep it tidy! It is also inefficient, as you may spend time looking for things you've mislaid.

The importance of organisation extends to your exercise books and revision folders. You will have at least two years' worth of work to revise and study. Misplacing work can have a negative effect on your revision.

Put together a revision folder with all your notes, practice questions, checklists, timetables and so on. You could organise it by topic so it's easy to find particular information and see the work you have already completed.

Looking after yourself

Revising for exams is a marathon, not a sprint – you don't want to burn out before you reach the exams. Make sure you stay healthy and happy while revising. This is important for your own wellbeing, and helps you revise effectively.

Eat properly

Try to eat a healthy, balanced diet. Keep some healthy snacks nearby so that hunger doesn't distract you when revising. Food high in sugar is not ideal for maintaining concentration, so make sure you're sensible when selecting snacks.

> **Tip**
>
> Some students find listening to music helpful when they're revising, even associating certain artists or songs with specific topics. However, music can also be distracting, so only use it if it works for you.

> **Tip**
>
> Be aware that social media can also be distracting. It's easy to procrastinate if you're not focused. Advice on reducing distractions can found on page 82.

Drink plenty of water

Make sure you have enough water at hand to last your revision session. It's vital to stay hydrated and getting up to get a drink can be a distraction, particularly if you wander past the TV on the way.

Consider when you work most effectively

Different people work better at different times of the day (morning, afternoon, early evening). Try to plan your revision during the times you're most productive. This may take some trial and error at the start of your revision.

Make sure you get enough sleep

Lack of sleep can lead to serious health problems. Late-night cramming is not an effective revision technique.

Avoiding distractions

Social media and other technology can provide an unwelcome temptation when studying. Possible solutions to this distraction include:

Plan specific online activities during study breaks

This could be social media time, videos or gaming. This can also give you something to look forward to while you're working. Be careful to ensure you stick to the allotted break time and don't fall into the trap of 'just one more' video or game.

Switch technology off

Switching the internet off can be the most powerful productivity tool. Turn off your phone and consider avoiding the internet whilst studying, only turning them back on at the end of the study session or during a break. This removes the temptation to constantly check your phone or messages. If you do need access to a device while studying, there are a number of blocker apps and services that can limit what you are able to access.

Tell your family and friends

Make sure people know that you are planning to study for a specific period of time. They'll understand why you may not be replying to their messages and they will help you by staying out of your way. This can also help with positive reinforcement, as you can talk to them afterwards about the successful outcomes of the revision session.

» Useful revision techniques

Many students start their GCSE studies with little idea of how to revise effectively. There are many effective revision techniques that are worth trying. And remember, revision is a skill that needs to be learnt and then practised. It may take time to get to grips with some of these strategies, but it will be worth it if you put the effort in.

Memory aids

Before we get onto the revision techniques themselves, here are some tips on how to memorise particularly complicated information. Look out for opportunities to put these techniques into action.

> **Tip**
>
> Even if you're more productive in the evening, you still need to go to bed early enough to get enough sleep.

Elaboration

Elaboration is where you ask new questions about what you have already learnt. In doing this you will begin to link ideas together and develop your holistic understanding of the subject. The more connections between topics your brain makes automatically, the easier you will find recalling the relevant information in the exam.

For example, if you have just consolidated your notes on the structure of the plant transport system you might challenge yourself to make a list of all the similarities and differences between the transport systems of plants and humans.

This is useful because, by answering this type of question, your brain will form links between the topics and strengthen your recall while also improving your understanding of both plant and human transport systems.

As part of elaboration you can try and link ideas to real world examples. These will develop your understanding and help you memorise key facts. For example, when revising polymer structure in Chemistry, you could relate this to examples of polymers and how they're used.

Key term

Holistic: When all parts of a subject are interconnected and best understood with reference to the subject as a whole.

Tip

It is helpful to use these types of questions to create linked mind maps showing the connections between topic areas.

Mnemonics

Mnemonics are memory aids that use patterns of words or ideas to help you memorise facts or information. The most common type is where you create a phrase using words whose first letters match the key word or idea or you are trying to learn.

For example, living things can be classified into these taxonomic levels:

- **K**ingdom
- **P**hylum
- **C**lass
- **O**rder
- **F**amily
- **G**enus
- **S**pecies.

A mnemonic to help remember the order of these levels might be:

King **P**hilip **C**ame **O**ver **F**or **G**reat **S**amosas

Other mnemonics include rhymes, short songs and unusual visual layouts of the information you're trying to remember.

Tip

When it comes to mnemonics, the sillier the phrase, the better – they tend to stick in your head better than everyday phrases.

Memory palace

Memory palace is a technique that memory specialists often use to remember huge amounts of information. In this technique, you imagine a place (this could be a palace, as in the name of the technique, but it could be your home or somewhere else you're familiar with), and in this location you place certain facts in certain rooms or areas. These facts should, ideally, be associated with wherever you place them, and always stay in the same location and appear in the same order.

You may also find it helpful to 'dress' each fact up in a visual way. For example, you might imagine the information 'gravitational acceleration is ~10 m/s^2' being 'dressed' as the apple that fell on Newton's head. You might then place this apple in the kitchen in your memory palace, 10 metres high on top of one of your cupboards.

Through the process of associating facts and their imaginary location, you are more likely to correctly recall the fact when you come to revisit the 'palace' and locations in your mind.

Make your revision active

In order to revise effectively, you have to actually *do something* with the information. In other words, the key to effective revision is to make it active. In contrast, simply re-reading your notes is passive and is fairly ineffective in helping people retain knowledge. You need to be actively thinking about the information you are revising. This increases the chance of you remembering it and also allows you to see links between different topic areas. Developing this kind of deep, holistic understanding of the course is key to getting top marks.

Different active techniques work for different people. Try a range of activities and see which one(s) work for you. Try not to stick to one activity when you revise; using a range of activities will help maintain your interest.

Key term

Active revision: Revision where you organise and use the material you are revising. This is in contrast to passive revision, which involves activities such as reading or copying notes where you are not engaging in active thought.

Retrieval practice

Retrieval practice usually involves the following steps.

Step 1 Consolidate your notes

Step 2 Test yourself

Step 3 Check your answers

Step 4 Repeat

Step 1: Consolidate your notes

Consolidating notes means taking information from your notes and presenting it in a different form. This can be as simple as just writing out the key points of a particular topic as bullet points on a separate piece of paper. However, more effective consolidation techniques involve taking this information and turning it into a table or diagram, or perhaps being more creative and turning them into mind maps or flash cards.

Bullet point notes

Here is an example of how you might consolidate bullet point notes from a chunk of existing text.

Original text

Ultrasound waves are inaudible to humans because of their very high frequency. These waves are partially reflected at a boundary between two different media. The time taken for the reflections to echo back to a detector can be used to determine how far away this boundary is, provided we know the speed of the waves in that medium. This allows ultrasound waves to be used for both medical and industrial imaging.

Seismic waves are produced by earthquakes. Seismic P-waves are longitudinal and they travel at different speeds through solids and liquids. Seismic S-waves are transverse, so they cannot travel through a liquid. P-waves and S-waves provide evidence for the structure and size of the Earth's core. The study of seismic waves provides evidence about parts of the Earth far below the surface.

Consolidated notes

- Ultrasound frequency > 20 000, so humans can't hear them
- Ultrasound reflects and the time it takes an echo to return can be used to find distance between target and source
- Ultrasound is used in medicine and industry to obtain images

- Two types of seismic wave in earthquakes: longitudinal P-waves and transverse S-waves
- P-waves can travel through solids and liquids, S-waves through solids only
- Both give information about interior structure of the Earth, e.g. size of the core

Flow diagrams

Flow diagrams are a great way to represent the steps of a process. They help you remember the steps in the right order. An example of a Chemistry flow diagram, for the Haber process, is shown in Figure 4.1.

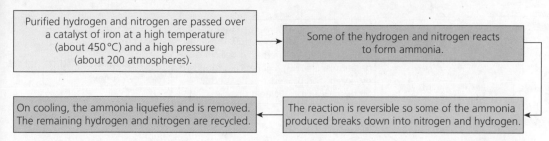

▲ Figure 4.1 **The Haber process**

Mind maps

Mind maps are summaries that show links between topics. Developing these links is a high-order skill – it is key to developing a full and deep understanding of the specification content.

Mind maps sometimes lack detail, so are most useful to make once you have studied the topics in greater detail.

See **Elaboration** (page 83) for more information on the importance of linking ideas in active revision.

> **Key term**
>
> High-order skill: A challenging skill that is difficult to master but has wide ranging benefits across subjects.

Good example of a mind map

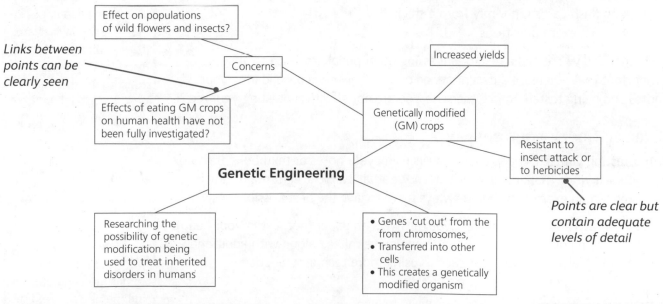

▲ Figure 4.2 **Example of a good mind map**

Bad example of a mind map

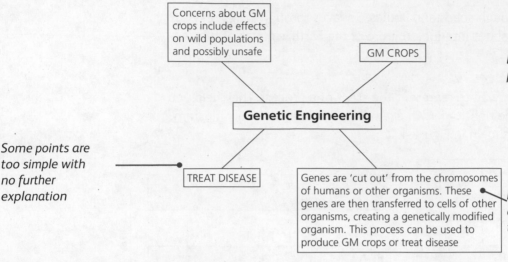

Concerns about GM crops include effects on wild populations and possibly unsafe

GM CROPS

No evidence of linking the points together

Genetic Engineering

Some points are too simple with no further explanation

TREAT DISEASE

Genes are 'cut out' from the chromosomes of humans or other organisms. These genes are then transferred to cells of other organisms, creating a genetically modified organism. This process can be used to produce GM crops or treat disease

Points like these are too complex and contain too much text

▲ Figure 4.3 Example of a bad mind map

Flash cards

Flash cards are excellent for things such as key word definitions – write a key word on one side of the card and the definition on the other.

Flash cards can also be used to summarise key points of a process or topic area.

Similar to mind maps, they should be used in conjunction with other revision methods that fully cover the detail required.

Step 2: Test yourself

There is a range of test activities you can do with the notes you've consolidated, including:

- making your own quizzes
- asking friends or family to test you
- picking flash cards randomly from a stack
- trying past exam questions

Whether you've created a test, or are asking other people to test you, it's important that you leave a decent period of time between consolidating your notes and being tested on it. Otherwise, you are not effectively testing your recall.

Step 3: Check your answers

After testing yourself, check your answers using your notes or textbooks. Be hard on yourself when marking answers. An answer that's *almost* right might not gain full credit in an exam. You should always strive to give the best possible answer.

If you get anything wrong, correct your answers on paper (not just in your head). And annotate your answers with anything you've missed along with additional things you could do to improve, such as using more technical language.

Step 4: Repeat

Repeat the whole process, for each topic, at regular intervals. Revisiting activities will help you memorise key aspects and ensure you learn from your previous mistakes. It is especially helpful for topics you find challenging.

> **Tip**
> Like with mind maps, do not squeeze too much information on a flash card.

> **Tip**
> There are a number of different apps that are useful to help with creating quizzes. Some of these apps also allow you to share these quizzes with friends, so you can help each other out.

> **Tip**
> Even though spacing out and mixing up topics are separate sub-sections here, they should be incorporated into your retrieval practice.

When repeating, do not *immediately* revisit the same topic again. Effective revision is more likely if you leave time before revisiting topics you've recently revised, and use this time to mix in other topics.

Spacing out topics

Once you've gone through a whole topic, move on and wait before returning to it and testing your recall. Ideally, you should return to a topic regularly, increasingly long intervals between each return. Returning to a topic needn't take too long – quickly redoing some tests you took before may be enough.

When you return, ask yourself:

- do you know the topic as well as you did when you revised it first time around?
- are you still making the same mistakes?
- what can you improve on?

Identify the key areas you need to go back over.

Allow time for this revisiting process in your revision timetable. Leaving things to the last minute and trying to cram is not an effective way of revising.

Tip
This is sometimes referred to as 'Spaced Practice'.

Mixing up topics

Mixing up topics (covering a mix of topics during your revision timetable rather than spending long periods of time on one) is an effective revision strategy. It ties into the need to revisit topics at intervals. Mixing up and revising different areas means it's inevitable there will be a space between first revising a topic and then coming back to it a later date.

Tip
This is sometimes referred to as 'Interleaving'.

Studies have shown that, although moving onto different topics more regularly may seem difficult, it could significantly improve your revision. So it's worth persevering.

➤➤ Practise, practise, practise

Completing practice questions, particularly exam-style questions, allows you to apply your knowledge and check that your revision is working. If you're spending lots of time revising but finding you cannot answer the exam questions, then something's wrong with your revision technique and you should try a different one. Examples of practice questions can be found on pages 104–108.

Practice exam questions can be approached in a number of ways.

Complete the questions using notes

This may seem a bit like cheating but it is good, active revision and will show you if there are any areas of your notes that need improving.

Complete questions on a particular topic

After revising a topic area, complete past exam questions on that topic without using your notes. If you find you get questions wrong, go back over your notes before returning to complete questions on this topic area again at a later date. Repeat this process until you are consistently answering all the questions correctly. Annotate your revision notes with points from the mark schemes. More details on the use of mark schemes can be found on pages 50–51.

Complete questions on a topic you have not yet revised fully

This will show you which areas of the topic you know already and which areas you need to work on. You can then revise the topic and go back and complete the question again to check that you have successfully plugged the gaps in your knowledge.

Complete questions under exam conditions

Towards the end of your revision, when you're comfortable with the topics, complete a range of questions under timed, exam conditions. This means in silence, with no distractions and without using any notes or textbooks.

It is important to complete at least some timed activities under exam conditions. The point of this is to prepare you for the exam. Remember, if you spend time looking up answers, talking, looking at your phone and so on, you won't get an accurate idea of timings.

Always ensure you leave enough time to check back over all your answers. Students often lose lots of marks due to silly mistakes, particularly in calculations. These can be avoided by ensuring you check all answers thoroughly.

When working your way up to completing an exam under timed conditions, it can be helpful to begin with timing one or two questions to get yourself used to the speed at which you should be answering them. You can then slowly work your way up to completing full-length papers in the time you would have in the real exam. Make a note of the areas where you found you were spending too long and look for ways to improve.

Effective revision is absolutely vital to success in GCSE Science. As you are studying a linear course you'll be examined on a whole two years' worth of learning. Only by revising effectively and thoroughly can you ensure you have a full and complete understanding of all the content.

> **Tips**
>
> As a guide to timings, you can work out how many marks you should be ideally gaining per minute. To do this, divide the total number of marks available by the time you have in the exam. This will help you get an idea of what questions need longer, but it is not a perfect guide as some questions will take longer than others, particularly the more complex questions that are often found towards the end of the exam paper.

5 Exam skills

Playing a musical instrument well takes a lot of practise; few musicians are immediately skilful, but over time their expertise develops. Learning a new musical piece takes effort and repeated rehearsing in order to improve. Preparing for your GCSE Chemistry examinations is a similar process. Even though the examinations for your GCSE Chemistry are only at the end, you should begin exam preparation almost immediately on beginning the two-year course. There are several areas that you should look at throughout the course to help you succeed in your exam.

» General exam advice

To perform to the best of your ability in the examination you need to consider the following points.

Before the exam
Exam specifics

It is very important that you look at the specification for the examination board you are studying. Each examination board has their specification online. The specification outlines the Chemistry content, which you will need to learn. You also need to note down:

- how many papers you will sit
- the length of time allocated for each paper
- what percentage of your GCSE is allocated to each paper
- the type of questions typically used – these may be multiple choice, linking answers, short answer, structured and extended response.

Practise answering full past papers to help you get used to the length of paper and examination style presented by your examination board. Study the mark schemes carefully and ensure that you note which marking points you did not include in your own answer. Mark schemes show exactly what the examiner is looking for and they will enable you to give a more detailed, precise and focused approach, which will help you improve your examination technique and answering style.

Planning ahead

The night before the examination, check the time of your paper. Sometimes students get the date of the examination fixed in their heads but mix up the timings – a critical mistake!

Arrive in good time for each examination; allow extra time for your journey in case of unforeseen delays.

Make sure you have everything you need including your calculator, ruler, pencil and pens. If you are using a pencil case, it must be transparent.

> **Tip**
> Write the number and length of each Chemistry paper on a post it and stick it near your desk to remind you of what you are working towards. A list of topics featured on each paper is also handy to have.

Avoid those students who might make you feel nervous before the examination. Sometimes, other students will say things to boost their own confidence; it is best not to get into last-minute discussions.

During the exam
Understanding what to do

When you are in the examination hall and are waiting to begin:

- check the instructions on the front of the paper carefully,
- check that you have the materials that you need, including the Data Booklet that is provided as an insert,
- ensure that you have filled in your details on the front cover.

At this stage, if you have practised past papers you should be familiar with the layout, style and timing of the paper and this should help you to feel more confident.

When you are waiting to start your exam, there is no point looking backwards over what you have not covered. Focus fully on the examination ahead and do your best to apply your knowledge to each question. During the exam, be focused and work hard. If you are struggling with a question, leave it out and move on.

Time management

The time allocation for your GCSE papers gives you about 1 minute per mark – try and stick to these timings. Some questions will require more time than others – especially the calculations. If you are struggling with a calculation or a part of a question, leave it out and move on. Then, if you have time at the end you can come back and complete it.

You should have enough time to answer all questions in the paper, but you need to ensure that you do not include irrelevant detail that may otherwise waste that time. Do not repeat the question when starting your answer. Remember that the key to success is obtaining maximum marks in minimum words.

Your question papers will start with a lower demand question and then slowly the questions will go up in difficulty. This should help build your confidence. Similarly, within a multipart question there will be an easier lead in, building through successive parts of the question. This means you have a fair chance of gaining some marks on each topic area throughout the paper. Don't give up when you come to a hard question, just move onto the next one as it will likely be easier. Draw a big star beside the question you left and come back to it at the end if there is time.

» Answering the question

The mark allocation given at the end of each question is very useful as you can use it to estimate the amount of detail to include in your answer. For example, a one-mark question would not require as much detail as a three-mark question. Avoid writing more than is necessary, for example, a one-mark question would not need more than one valid point, while a two-mark question would require you to make two points in your answer.

For example: *What is meant by the term 'weak acid'?*   [2]

Tip

Each time you turn a page, check the page number you have turned to is the next sequential one, and put a line through it. This will help you to avoid missing out pages.

Tip

A structured question gets progressively more difficult as you work through it. If you get stuck on a harder part, move on to the next question (it will start with an easy part on a different topic) and build your confidence again.

Tip

Don't get stuck on a part of the question that might only be worth one mark, try to move onto the next part if possible.

As this is a two-mark question, you should make two points in your answer such as,

- a weak acid is partially ionised
- in aqueous solution.

Some parts of a question may be highlighted in **bold print**. Bold is used to help you focus on the answer. For example, a question that states 'Give **one** reason' means do not give more than one reason; if you do, you are likely to lose marks.

Read each question very carefully and make sure that you answer what was asked, it is sometimes useful to underline or highlight parts of the question to ensure you are focused on the key words. Key words are the important words in the question that help you to explain things. Key words include names of objects or substances such as 'hydrogen' or 'flask', the names of processes such as 'evaporation' or 'distillation', and concepts such as the idea of bonding. Always check you are referring to or explaining the key words in the question. For example, a question may be about the atomic structure of a potassium ion, but it is easy to overlook the word 'ion' and incorrectly write about the potassium atom. Underlining key terms will focus your attention.

> **Key term**
>
> Key word: A word that helps you communicate ideas in science clearly.

Diagrams, graphs or tables, if provided in the question, may help you answer the question. Always study these carefully and think about what the information might mean. For example, ask yourself:

- what are the units in the table or graph?
- how does the diagram (table or graph) link to the information given?
- how can I use the information to help answer the question?
- what trends are there?

Showing your working

For all calculations you should show your working out. Remember that if you get the answer wrong but have shown working you can gain marks for the correct working. On that basis, it is good practice to show working to ensure that if you make an error it is still possible for you to score some marks.

For instance, the question '*Calculate the formula mass of $Ca(OH)_2$*' would typically be worth two marks. If a student writes down just the final and incorrect answer such as '*93*' they will not score any marks. However, if they had provided some working to show how they had come to this answer, they may have been able to gain at least one mark if that working had been partially correct. For instance, writing down '*40+(16+1) x 2=93*', for this question will score at least one mark as the calculation is correct in theory, even though the final answer is wrong. The correct answer to this question is 74.

Other common troublemakers

To help you gain maximum marks on your examination paper there are certain areas where you need to be very careful when presenting your answer.

Chemical symbols and formula

Remember that symbols have a capital, followed by a lower-case letter. The use of incorrect symbols will be penalised on your examination paper. For example, the use of 'h' for hydrogen, 'CL' for chlorine or 'br' for bromine will all be penalised. Incorrect formulae will also be penalised, for instance, as Na_2CO_3 is the correct formula for sodium carbonate, even a small error such as labelling it $NaCO_3$ will not be credited.

Equations

Always check that your equations include all the correct formula and are correctly balanced.

Organic structure

When drawing a structural formula make sure that each atom is bonded correctly. For example, for an alcohol formula, C–HO is incorrect; it must be written as C–OH.

» Assessment objectives

Every question on your examination paper tests one or more Assessment Objective (AO). Each Assessment Objective is split into several elements as shown in the table.

Assessment objective	Meaning	Approximate weighting (%)
AO1	Demonstrate knowledge and understanding of: scientific ideas; scientific techniques and procedures.	40
AO2	Apply knowledge and understanding of: scientific ideas; scientific enquiry, techniques and procedures.	40
AO3	Analyse information and ideas to: interpret and evaluate; make judgements and draw conclusions; develop and improve experimental procedures.	20

AO1 questions

AO1 questions aim to assess your knowledge and understanding of both theory from the specification and the core practicals. Each AO1 question is usually worth only a small amount of marks unless you are being asked to recall a lot of separate facts. A typical AO1 question might be:

A Worked example

1 Name the ore from which aluminium oxide is made.   [1]

Model answer

Bauxite

The answer is a fact that you should have learnt from your specification.

AO2 questions

Recall of facts is not enough to obtain a good grade; many questions require you to apply your knowledge to different and unfamiliar contexts. AO2 questions aim to assess your ability to apply scientific ideas, theories, scientific enquiry, practical skills and techniques, to explain phenomena and observations in familiar and unfamiliar contexts. Often these questions are set in novel theoretical and practical contexts.

In terms of application of practical knowledge this could be application of a technique or procedure to a novel situation. It could also be application of investigative skills, for example, data analysis. Maths, including graphs and chemical equations, is also assessed under AO2.

A **Worked example**

1 Calculate the mass of sodium hydroxide required to make 1000 cm³ of a
   0.25 mol/dm³ solution of sodium hydroxide.   [2]

Model answer

M_r NaOH $= 23 + 16 + 1 = 40$

0.25 mol/dm³ means there are 0.25 mol in 1000 cm³

Moles of NaOH $= 0.25 = \dfrac{\text{mass}}{M_r} = \dfrac{\text{mass}}{40}$

Mass $= 40 \times 0.25 = 10\,\text{g}$

Tip
· · · · · · · · · ·
Most questions that
ask you to write
balanced symbol
equations, or draw
bonding diagrams
for covalent or ionic
bonding, are also
AO2 as they require
you to apply your
knowledge of the
topic to a specific
chemical.

AO3 questions

AO3 questions require you to analyse information and use that to interpret
and evaluate, or draw conclusions, using your understanding of the underlying
science. As there is a requirement for analysis of information, questions will have
a stimulus for you to work, for example, from a graph, photo or table.

A **Worked example**

1 A chromatography experiment was set up as shown below. Explain two errors that have been made in
   setting up this experiment.   [2]

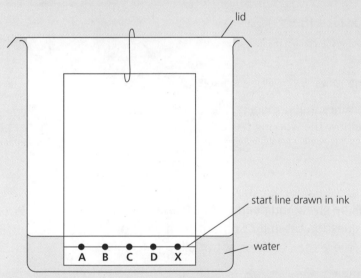

Model Answer

The start line with inks on it is below the level of water. This will cause the inks to dissolve. The start line
should be above the level of water, so the water moves up over it.

The start line is drawn in ink and the ink will dissolve and ruin the results of the experiment. The start line
should be drawn in pen.

» Command words

Command words are the words and phrases used in exams that tell students how they should answer a question. Each command word is part of a command sentence, such as '**Explain** how sodium chloride conducts electricity'. The command word almost always occurs at the start of the sentence.

You should always underline the command word in the question and focus on it before you start your answer. It is very easy to lose marks by not doing what the question tells you to do.

Tip

For the full list of command words and their definition, see page 123.

Command word: 'Calculate'

In a 'Calculate' question, you should use numbers given in the question to work out the answer. A numerical answer is expected. You may also be asked to include the correct units with your answer or to write it to a certain number of significant figures. Sometimes, you will need to choose the correct equation to use and substitute the correct numbers into the equation to obtain your answer.

A Worked example

1   Calculate the relative formula mass of magnesium nitrate $Mg(NO_3)_2$.   [2]

(relative atomic masses: $Mg = 24$, $N = 14$, $O = 16$)

Model answer

$$Relative\ formula\ mass = 1 \times Mg + (2 \times N) + (6 \times O)$$
$$= 1 \times 24 + (2 \times 14) + (6 \times 16) = 148$$

Note that for relative formula mass there are no units.

This is a model answer because it correctly identifies the formula needed, substitutes in the correct values and completes the calculation accurately. Note how the working has been shown to maximise the chances of gaining marks.

Command word: 'Choose'

In a 'Choose' question, a list of alternatives will be given. You need to select the correct one to answer the question. Read the question carefully. Sometimes it will state that each answer may be used once, more than once or not at all.

A Worked example

1   Some covalent substances, A–D, are shown below.

Choose the substance, A, B, C or D,

i   which represents methane   [1]
ii which represents a diatomic element.   [1]

Model answer

> i   Methane has the formula CH_4. The answer is A.
>
> ii Only D is an element, made up of one type of atom, and it is diatomic as there are two atoms covalently bonded in the molecule. The answer is D.

This is a model answer because it correctly chooses the information needed from the list. No other information is needed.

Command word: 'Compare'

In a 'Compare' question, you need to describe the similarities and/or differences between things. The key to answering 'Compare' questions is to ensure that you include comparative statements. See pages 48–49 for guidance on how to answer this type of question.

Command word: 'Complete'

In a question that asks you to 'Complete', you need to write your answers in the space provided. This may be on a diagram, within blank spaces in a sentence or in a table.

(A) Worked example

Complete the diagram below to show the electronic configuration of magnesium. [1]

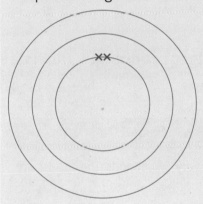

Model Answer

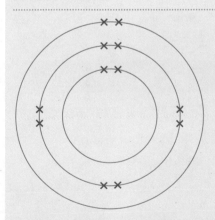

> **Tip**
> Magnesium has atomic number of 12 and so 12 electrons are arranged as 2,8,2.

This is a model answer because it completes the diagram that was started, and includes the correct amount of detail and information required by the question.

Command word: 'Define'

When a question asks you to 'Define', it means you have to state the meaning of a term. Keep your answer simple and to the point. It is a good idea to make a list of all the definitions given within your specification.

A Worked example

1 Define the term carbon footprint. [1]

Model Answer

The carbon footprint is the total amount of carbon dioxide and other greenhouse gases emitted over the full life cycle of a product, service or event.

This is a model answer because it provides a concise definition, including correct use and spelling of technical terms.

Command word: 'Describe'

Questions that ask you to 'Describe' want you to give a detailed account, in words, of relevant facts and features relating to the topic being examined. For guidance on how to answer this type of question see pages 51–53.

Command word: 'Design'

Questions that ask you to 'Design' want you to set out a procedure showing how something can be done. For guidance on how to answer this type of question see page 56.

Command word: 'Determine'

This type of answer requires you to use information or data, given for example in tables or graphs, to obtain the answer.

A Worked example

1 In an experiment a student recorded the mass of a flask and its contents every 50 seconds. The results are shown in the following graph. Determine the time taken for the reaction to finish. [1]

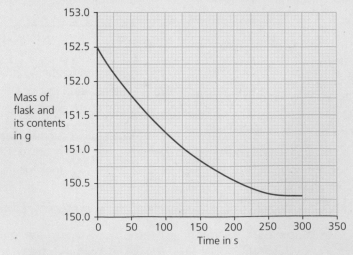

Model answer

The graph must be used to find the time when the mass of the flask and its contents stays constant and the reaction is over. This is at 260 seconds.

This is a model answer because it correctly determines the data that you are asked to identify. The units are also included in the answer. Some 'Determine' questions might also require mathematical calculation.

Command word: 'Draw'

You may be asked to draw a complete diagram or add to a diagram for processes such as filtration or evaporation.

A Worked example

Propene is used to produce poly(propene). Draw the bonds to complete the displayed formula of propene and poly(propene) in the equation below. [2]

$$n \quad \begin{matrix} H & H \\ C & C \\ H & CH_3 \end{matrix} \longrightarrow \left(\begin{matrix} H & H \\ C & C \\ H & CH_3 \end{matrix} \right)_n$$

Model answer

Remember that a polymer does not have a double bond and the bonds project out through the brackets to show that it repeats.

$$n \quad \begin{matrix} H & H \\ | & | \\ C{=}C \\ | & | \\ H & CH_3 \end{matrix} \longrightarrow \left(\begin{matrix} H & H \\ | & | \\ C{-}C \\ | & | \\ H & CH_3 \end{matrix} \right)_n$$

This is a model answer because it is a clear diagram with all the atoms correctly joined using bonds and the correct number of bonds.

Command word: 'Estimate'

'Estimate' means to give an approximate amount. See page 17 for more information on giving approximate amounts.

A Worked example

1 The boiling points of some of the halogens are shown in the table.

Halogen	Boiling point (°C)
fluorine	−188
chlorine	
bromine	60
iodine	184

Estimate the boiling point of chlorine. [1]

Model answer

The boiling point of chlorine will be in between that of fluorine and chlorine. Choose a number midway, for example -50°C.

This is a model answer because it provides an answer in the rough ballpark. You don't need to have an exact answer for 'Estimate' questions, although you won't be penalised for using exact figures; but, given the short period of time usually allowed for these types of questions, you are unlikely to have enough time for the full calculation.

Command word: 'Evaluate'

In an 'Evaluate' question, you should use information supplied in the question, and your own knowledge, to consider evidence for and against. This command word will usually be used in longer answer questions, and you should ensure that you give points both for and against the idea you have been asked to evaluate. For guidance on how to answer this type of question see pages 57–59.

Command word: 'Explain'

Questions that ask you to 'Explain' want you to make something clear, or state the reasons for something happening. For guidance on how to answer this type of question see pages 54–55.

Command word: 'Give'

Only a short answer is required in these types of questions. An explanation or a description is not needed.

> **Tip**
> Note the difference between 'Explain' and 'Describe'; 'Explain' is *why* something is happening while 'Describe' is *what* is happening.

(A) Worked example

A sample of water was tested to determine if chloride ions were present. A few drops of nitic acid were added followed by a few drops of a reagent X. Give the name of the reagent X. [1]

Model answer

Silver nitrate

This is a simple model answer that answers the question – you do not need to provide any additional information or reasoning unless specified by the question (or if you suspect there are more marks available than you would normally expect for just stating one idea).

Command word: 'Identify'

Questions that ask you to 'Identify' want you to select key information from a source provided for you. Other 'Identify' questions might be more complex than just choosing from the source. For example, questions like 'Identify the independent variable' would require you to recall what an independent variable is and apply it to the information provided. Even so, your answer for any 'Identify' question can be rather short and direct.

> **Tip**
> A correct name or a formula is often an appropriate answer to an 'Identify' question.

(A) Worked example

1 Identify the substance that is oxidised in the reaction:

$$Fe_2O_3 + 3CO \rightarrow 2Fe + 3CO_2$$ [1]

Model answer

CO/Carbon monoxide

This is a model answer because, like 'Choose', you are only required to pick the correct answer from the options.

Command word: 'Justify'

When you answer a 'Justify' question you must select evidence from the information given in the question and use it to support your answer.

(A) Worked example

1 A few drops of sodium hydroxide solution were added to a solution of a metal compound A and a white precipitate was formed that dissolved when excess sodium hydroxide solution was added. A few drops of nitric acid were then added to separate solution of A, followed by a few drops of silver nitrate. A white precipitate was produced.

Identify compound A. Justify your answer. [3]

Model answer

'A' is aluminium chloride. [1]

This is because solutions of magnesium, calcium and aluminium ions form white precipitates with sodium hydroxide but only aluminium hydroxide precipitate dissolves when excess sodium hydroxide solution is added. [1]

The negative ion is chloride as it reacts with silver nitrate to form a white precipitate of silver chloride. [1]

This is a model answer as it not only correctly identifies the correct option but crucially it also gives a well-reasoned argument to support it.

Command word: 'Label'

When asked to 'Label' you should provide appropriate names on a diagram, illustration or graph to indicate which particular item they are pertaining to.

(A) Worked example

Label the two pieces of apparatus indicated in the diagram below. [2]

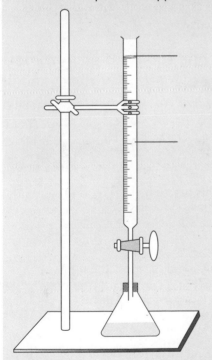

Model answer

Burette [for the top label]

Conical flask [for the bottom label]

This is a model answer because all the labels are completed correctly. Usually the labels you have to complete will be signposted as they are in this question, but on rare occasions you may have to draw your own. In those cases, make sure it's really clear what the labels are pointing at.

Command word: 'Measure'

Questions on examination papers that ask you to 'Measure' may require you to 'Measure' a distance using a ruler.

A Worked example

1 Measure the distance moved by the solvent and the distance moved by the blue dye in the chromatogram below.

[2]

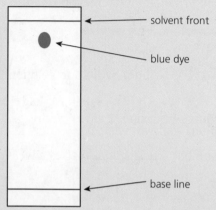

solvent front

blue dye

base line

Model answer

In chromatography, the measurement is always taken from the base line and to the middle of the spot.

Distance moved by solvent = 4.2 cm

Distance moved by blue spot = 3.7 cm

This is a model answer because it correctly determines the data that you are asked to work out and includes a unit. In this example, a ruler also has to be used correctly to arrive at the answer.

Command word: 'Name'

In a 'Name' question, only a short answer is required. Do not be tempted to give an explanation or a description. Often, you may only need to use a single word, phrase or sentence.

A Worked example

1 Calcium carbonate reacts with hydrochloric acid. Name the products in this reaction.

[1]

Model answer

Calcium chloride. Carbon dioxide. Water

This is a model answer because, like 'Identify' questions, you are simply being asked to select the right information and write it down (either from your own knowledge or from information provided). You only need to include the words you are being asked to 'Name', do not waste time writing sentences.

Command word: 'Plan'

Questions that ask you to 'Plan' want you to give detailed information about how a procedure or task might be carried out. For guidance on how to answer this type of question see pages 55–7.

Command word: 'Plot'

Questions that ask you to 'Plot' want you to draw and label axes on a grid and mark the points provided. If there is a correlation, you may also be asked to draw the line(s) of best fit. For guidance on how to answer this type of question see pages 33–37.

Command word: 'Predict'

This question expects you to give a plausible outcome. Data are often given in this type of question and you should study it to work out any trends that help with your prediction.

A Worked example

1 The formulas of three different alkanes are shown in the table.

Name	Formula
ethane	C_2H_6
propane	C_3H_8
butane	C_4H_{10}

Predict the formula of the alkane that contains 5 carbon atoms. [1]

Model answer

The table shows a pattern in the formula. The alkane with 5 carbon atoms is C_5H_{12}

This is a model answer because it provides a plausible outcome from analysing the information given. You do not have to provide the reasoning for your prediction unless the question asks for it.

Command word: 'Show'

Questions that ask you to show need you to provide structured evidence to reach a conclusion. This may involve carrying out a calculation.

A Worked example

1 In a reaction 0.06 g of magnesium reacts with excess hydrochloric acid, and the hydrogen produced is collected in a gas syringe. The equation for the reaction is

$$Mg + 2HCl \rightarrow MgCl_2 + H_2$$

Show that a 100 cm^3 gas syringe can be used to collect the hydrogen. [3]

Model answer

$$\text{Moles Mg} = \frac{0.06}{24} = 0.0025 \qquad [1]$$

1 mole Mg : 1 mole H_2

$$\text{Moles } H_2 = 0.0025 = \frac{vol(cm^3)}{24\,000} \qquad [1]$$

Volume = 60 cm^3 = 24 000 × 0.0025 = 60 cm^3, which is less than 100 cm^3, so this size of gas syringe can be used. [1]

This is a model answer because it 'Shows' the correct calculations and facts, that is, suitable evidence, needed to get to the right answer given in the question.

Command word: 'Sketch'

If the question asks you to 'Sketch' a line, or a graph for example, it means you should draw it approximately. Do not plot exact points.

A Worked example

1 A student reacted a strip of magnesium ribbon with solutions of hydrochloric acid of three different concentrations (0.5, 1.0 and 1.5 mol/dm³) at room temperature. The results obtained are shown on the graph below.

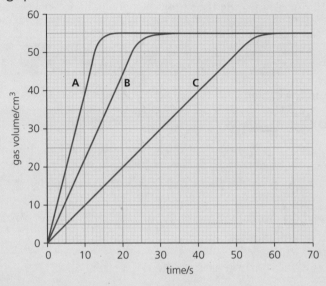

The student repeated the experiment using hydrochloric acid at a concentration of 2.0 mol/dm³. Sketch a line on the same axes to represent the results obtained. [1]

Model answer

The 2.0 mol/dm³ acid is more concentrated, so the reaction will be faster. This is shown in the sketch by a line that is steeper than the 1.5 mol/dm³, which finished earlier but finishes at the same volume of gas.

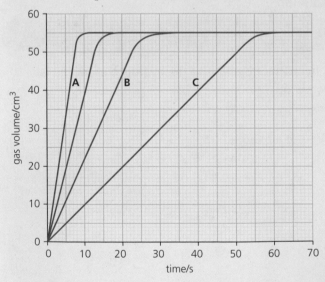

This is a model answer because it clearly shows all the information requested by the question in a clear fashion. Only the general shape of the graph is required for 'Sketch'.

Command word: 'Suggest'

This term is used in questions where candidates need to apply their knowledge and understanding to a new situation. Often, there may be more than one correct answer as candidates are expected to base their answers on scientific knowledge.

A Worked example

1 A student carried out some displacement reactions to put the metals zinc, iron and copper in order of reactivity. He added samples of the metal to a measured volume of copper(II) sulfate solution and recorded the temperature. Suggest why the student would not use potassium in this experiment.   [1]

Model answer

Potassium is a very reactive metal and it is dangerous to react it with copper(II) sulfate solution.

This is a model answer because it addresses the points in the question. The suggested explanation based on the scenario given is a plausible one, backed up by scientific knowledge, which is clearly explained.

Command word: 'Use'

In a 'Use' question, the answer must be based on the information given in the question. For guidance on how to answer this type of question see pages 60-2.

Command word: 'Write'

In this case a short answer is required, not an explanation or a description. Often the question can be answered with a single word, phrase or sentence. Read the question carefully as sometimes you may be asked to 'Write' down one (or two etc.) examples. You should only 'Write' down the specified number of answers as you may lose marks for any wrong examples given.

A Worked example

1 Write down the symbols of two elements that are liquids at room temperature and pressure.   [2]

Model answer

Br, Hg

This is a model answer because it answers the question in a succinct and clear manner. For 'Write' questions, you should not include any more information than is required.

Put this into action

Now that you know what all the main command words mean and how to answer them, the next and most important step is to put this learning into action. The next section provides some exam-style practice questions for you to apply your knowledge and help you prepare for the exam. Don't forget that there are also past and sample assessment materials for your specific exam board online.

6 Exam-style questions

» Paper 1

1 A student carried out an experiment to determine if the reaction between hydrochloric acid and sodium hydroxide was exothermic. The student followed the method here.

- Measure out 25.0 cm³ of 0.10 mol/dm³ hydrochloric acid and place in a polystyrene cup.

- Record the temperature of the hydrochloric acid.

- Gradually add 25.0 cm³ of sodium hydroxide solution in 5.0 cm³ portions to the hydrochloric acid, stirring after each addition.

- Record the temperature of the reaction mixture.

The table shows the student's results.

Volume of sodium hydroxide added in cm³	0.0	5.0	10.0	15.0	20.0	25.0
Temperature of reaction mixture in °C	20.5	21.5	22.5	23.5	25.2	28.0

    a Plot a graph of the results. Use axes similar to those shown below.   [3]

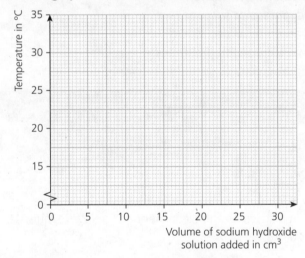

    b State why your graph shows that this reaction was exothermic.   [1]

    c Name a piece of apparatus that could be used to add the sodium hydroxide solution to the acid.   [1]

    d Suggest one improvement that could be made to the apparatus used that would give more accurate results. Give a reason for your answer.   [2]

    e Write a balanced chemical equation for the reaction between sodium hydroxide and hydrochloric acid.   [2]

    f Calculate the number of moles of hydrochloric acid placed in the polystyrene cup.   [1]

    g Hydrochloric acid is a strong acid. What is meant by strong acid?   [1]

    h A student wished to change the experiment to determine the temperature at neutralisation. Suggest one change that could be made to the experiment.   [2]

i The sodium hydroxide solution used was made by dissolving 40.0 g of sodium hydroxide in water and making the solution up to 250 cm^3 with water. Calculate the concentration of the solution in mol/dm^3. [2]

2 Bath crystals contain Epsom salts, which are hydrated magnesium sulfate crystals. Magnesium sulfate crystals can be prepared in the laboratory by reacting magnesium carbonate and sulfuric acid. The equation for the reaction is:

$$MgCO_3 + H_2SO_4 \rightarrow MgSO_4 + H_2O + CO_2$$

a State what could be is observed in this reaction. [1]

b Describe how a sample of magnesium sulfate crystals could be made from magnesium carbonate and dilute sulfuric acid. [6]

c Suggest one safety precaution that should be followed. [1]

d Calculate the maximum mass of magnesium sulfate that could be made when 2.1 g of magnesium carbonate is reacted with excess sulfuric acid. [3]

e The student obtained 1.8 g of magnesium sulfate. Calculate the percentage yield. [2]

f Suggest one reason why the percentage yield is not 100% in this reaction. [1]

3 A student placed 25.0 cm^3 of white wine, containing tartaric acid, in a conical flask. The student carried out a titration to find the volume of 0.100 mol/dm^3 sodium hydroxide solution needed to neutralise the tartaric acid in the white wine.

a Name a suitable indicator for this titration and the colour change that would be seen. [2]

b Suggest why this titration is suitable for white wine, but it is not used to find the concentration of acid in red wine. [1]

c The student carried out four titrations. Her results are shown in the table. Concordant results are within 0.10 cm^3 of each other.

	Volume of 0.100 mol/dm^3 NaOH in cm^3
Titration 1	20.05
Titration 2	19.45
Titration 3	18.90
Titration 4	19.00

i Use the student's concordant results to calculate the mean volume of 0.100 mol/dm^3 sodium hydroxide added. [2]

The equation for the reaction of tartaric acid in the white wine with the sodium hydroxide is:

$$C_4H_6O_6 + 2NaOH \rightarrow C_4H_4O_6Na_2 + 2H_2O$$

ii Calculate the concentration, in mol/dm^3, of the tartaric acid. Give your answer to two significant figures. [5]

iii Calculate the relative formula mass of tartaric acid ($C_4H_6O_6$) [1]

4 A student decides to investigate the reactivity of the four metals, P, Q, R and S. Write a plan for how the student could investigate the relative reactivity of the four metals, P, Q, R and S. The plan should use the fact that all four metals react exothermically with dilute sulfuric acid. You should name any apparatus you will use. [6]

5 A student investigated the reaction of 0.1 g of magnesium ribbon with 50 cm^3 of dilute hydrochloric acid of concentration 1 mol/dm^3 at 20 °C. The folllowing diagram shows the apparatus used.

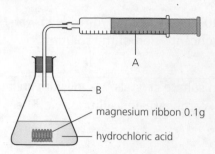

magnesium ribbon 0.1g

hydrochloric acid

a Name the pieces of apparatus A and B. [2]

b Complete and balance the following equation for the reaction between magnesium and hydrochloric acid.

..............................+............................ → ...H$_2$+.......................... [2]

c Give one advantage and one disadvantage of using a measuring cylinder to add the acid to the flask. [2]

d The table shows the results of this experiment.

Time in s	0	30	60	90	120	150	180	210
Volume of gas in cm^3	0	13	22	30	36	43	49	49

On the grid:

- plot these results

- draw a line of best fit. [3]

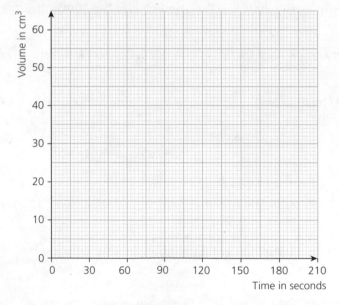

e Using your graph, find the time needed to collect 25 cm^3 of gas. [1]

f Calculate the mean rate of the reaction for the first 30 seconds of the reaction. Give your answer to one significant figure. State the units. [4]

g Using your graph, determine the rate of reaction at 60 seconds. Show your working on the graph. Give your answer in standard form. [4]

h Sketch a line on the grid on your graph to show the results you would expect if the experiment were repeated using 0.1g of magnesium filings in 50 cm^3 of 1 mol/dm^3 hydrochloric acid at 20 °C. Label this line A. [2]

i Explain how and why the rate of reaction would change if the experiment was repeated at 40 °C. [2]

6 Solder is an alloy of tin and lead.

  a A sample of a solder was made by mixing 22.5 g of lead with 15.0 g of tin. Calculate the percentage of tin by mass in this solder. [2]

  b Why are alloys stronger than pure metals? Pick one option from below. [1]

    A There are stronger bonds between the molecules they contain.

    B They combine the properties of the metals from which they are made.

    C They have atoms of different sizes in their structures.

    D They are made using electrolysis.

7 The diagram shows the results of a chromatography experiment to analyse a dye X.

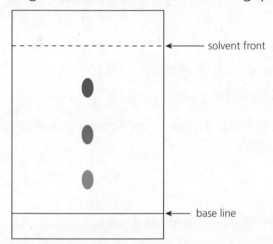

  a Explain why the base line must be drawn in pencil instead of pen. [1]

  b Identify how many dyes were in the mixture X. [1]

  c Calculate the R_f value for the blue spot. Use the table to identify it. [2]

dye	R_f value
A	0.38
B	0.15
C	0.26
D	0.75
E	0.58

8 New ways of extracting copper include phytomining and bioleaching. Describe the methods of phytomining and bioleaching and explain how copper metal is produced using these methods and scrap iron. [6]

9 There is less carbon dioxide in the Earth's atmosphere now than there was in the Earth's early atmosphere.

The amount of carbon dioxide in the Earth's early atmosphere decreased because plants and algae used it for photosynthesis and it became locked up in sedimentary rocks.

  a Photosynthesis can be represented by the equation shown.

    Complete the equation by writing the formula of the other product and balancing it correctly. [2]

    ....CO_2 + ... $H_2O \rightarrow$ .... + ... O_2

  b Explain what is meant by 'locked up carbon dioxide'. [2]

  c The graph shows how the amount of carbon dioxide in the atmosphere has changed in recent years.

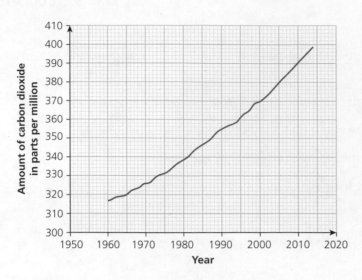

i   Describe how the amount of carbon dioxide has changed between 1960 and 2010.   [2]

ii   Calculate the percentage change in carbon dioxide levels between 2000 and 2010.   [2]

iii Give two reasons why the amount of carbon dioxide has changed over time.   [2]

10 Atmosphere is the term used to describe the collection of gases that surround a planet. The suggested composition of the atmosphere of Mars is shown in the table.

Gas	Composition (%)
Carbon dioxide	95.0
Nitrogen	3.0
Noble gases	1.6
Oxygen	trace
Methane	trace

Compare the composition of the Earth's atmosphere today with that of the planet Mars.   [4]

(Total marks ____ / 95)

For further practice, an extra paper is available online at
www.hoddereducation.co.uk/EssentialSkillsChemistry

Answers

» Maths

Units and abbreviations

Converting between units (page 3)

Guided questions

1   $1.2 \times 1000 = 1200\,cm^3$

2   **Step 1** $8.2 \times 1000 = 8200\,kg$

   Step 2 $8200 \times 1000 = 8\,200\,000\,g$

Practice questions

3   a)   To convert from dm^3 to cm^3 multiply by 1000

      $1.2 \times 1000 = 1200\,cm^3$

   b)   To convert from cm^3 to dm^3 divide by 1000

      $\dfrac{420}{1000} = 0.42\,dm^3$

   c)   To convert from cm^3 to dm^3 divide by 1000

      $\dfrac{3452}{1000} = 3.452\,dm^3$

   d)   To convert from tonnes to grams, first convert to kg by multiplying by 1000 and then convert to grams by multiplying by 1000 (or multiply by 10^6 in one step)

      $4.4 \times 1000 \times 1000 = 4400\,000\,g$

   e)   To convert from kg to g multiply by 1000

      $4 \times 1000 = 4000\,g$

   f)   To convert from g to kg divide by 1000

      $\dfrac{3512}{1000} = 3.512\,kg$

Making calculations involving conversion of units (pages 4–5)

Guided questions

1   **Step 1** $9.8 \times 1000 = 9800\,g$

   Step 2 Amount (in moles) $= \dfrac{mass\,(g)}{M_r} = \dfrac{9800}{98} = 100\,mol$

2   **Step 1** $\dfrac{48\,000}{1000} = 48\,dm^3$

   Step 2 Substitute the volume into the equation and calculate your final answer.

   amount (in moles) $= \dfrac{volume\,(dm^3)}{24} = \dfrac{48}{24} = 2\,mol$

Practice questions

3   The mass of calcium must be converted from kg to g before calculating moles.

   $6 \times 1000 = 6000\,g$

   amount (in moles) $= \dfrac{mass\,(g)}{A_r} = \dfrac{6000}{40} = 150\,mol$

4   The mass of calcium carbonate must be converted from tonnes to grams before calculating moles.

   $3.2 \times 1000 \times 1000 = 3\,200\,000\,g$

   amount (in moles) $= \dfrac{mass\,(g)}{M_r} = \dfrac{3\,200\,000}{100} = 32\,000\,mol$

5   The mass of ammonia must be converted from kg to g by multiplying by 1000.

   mass of ammonia in grams $= 17 \times 1000 = 17\,000\,g$

   Calculate the relative formula mass of $NH_3 = 14 + (3 \times 1) = 17$

   amount (in moles) $= \dfrac{mass\,(g)}{M_r} = \dfrac{17\,000}{17} = 1000\,mol$

6   Mass of iron(III) oxide in grams $= 2.1$ tonnes $\times 1000 \times 1000 = 2100\,000\,g$

   Relative formula mass $= (2 \times 56) + (3 \times 16) = 160$

   amount (in moles) $= \dfrac{mass\,(g)}{M_r} = \dfrac{2100\,000}{160} = 13\,125\,mol$

7   Mass of magnesium nitrate in grams $= 0.592 \times 1000 = 592\,g$

   Relative formula mass $= 24 + (2 \times 14) + (6 \times 16) = 148$

   amount (in moles) $= \dfrac{mass\,(g)}{M_r} = \dfrac{592}{148} = 4\,mol$

8   Volume of sulfur trioixide in $dm^3 = \dfrac{7200}{1000} = 7.2$

   amount (in moles) $= \dfrac{volume\,(dm^3)}{24} = \dfrac{7.2}{24} = 0.3\,mol$

Arithmetic and numerical computation

Expressions in decimal form (pages 6–7)

Guided questions

1   **Step 1** $\underline{3.41}8\,g$

   Step 2 The number after the underlined number is 8 so the rule 'if the next number is 5 or more, round up' is followed.

   The answer is 3.42 g (to 2 d.p.)

2   a)   **Step 2** $\underline{1.54}9\,g$

      Step 3 Round up 1.55 g

   b)   **Step 1** $28.465 - 27.799 = \underline{0.666}\,g$

      Step 2 0.666 g

      Step 3 The next number is 6 so the rule 'if the next number is 5 or more round up' is applied so the answer is 0.67 g

Practice questions

3   a)   1.72 g, which is 1.7 g (to 1 d.p.)

   b)   9.69 g, which is 9.7 g (to 1 d.p.)

   c)   2.11 g, which is 2.1 g (to 1 d.p.)

4   The cathode increases in mass and so copper is deposited here.
Mass $= 1.87 - 1.58 = 0.29\,g = 0.3$ (to 1 d.p.)

5

Mass /g	Mass recorded to 2 decimal places /g
29.883	29.88
0.046	0.05
32.6789	32.68
13.999	14.00
0.0894	0.09
19 992.456	19 992.46

Recording to an appropriate number of decimal places (pages 8–9)

Guided question

1   Step 1

Temperature /°C	Number of decimal places
10.25	2
10	0
10.2	1
10.0	1

Step 2 The answer should be recorded to 0 decimal places

Step 3 average $= \dfrac{10.25 + 10 + 10.2 + 10.0}{4} = 10.1125$

$= 10\,°C$ (to 1 d.p.)

Practice questions

2   a)   53.667 g is the total. The number with the least decimal places is 43.2 g with 1 decimal place, hence the answer must be rounded to one decimal place – 53.7 g.

b)   13.128 g is the total. The number with the least decimal places is 2.49 g with 2 decimal places, hence the answer must be rounded to 2 decimal places – 13.13 g.

c)   7.5439 g is the total. The number with the least decimal places is 3.23 g or 3.97 g with 2 decimal places, hence the answer must be rounded to 2 decimal places: 7.54 g.

3

Mass of evaporating basin and solid /g	Mass of evaporating basin /g	Mass of solid (to appropriate number of decimal places) /g
34.567	23.4	11.2 (to 1 d.p.)
29.93	25.66	4.27 (to 2 d.p.)
25.49	22.1	3.4 (to 1 d.p.)

Expressions in standard form (pages 10–11)

Guided questions

1   Step 1 6.45×10^n

Step 2 6.45×10^{-3}

2   Step 1 Amount (in moles) $= \dfrac{\text{mass}(g)}{M_r} = \dfrac{2.3}{23} = 0.1$

Step 2 $0.1 \times 6.02 \times 10^{23} = 6.02 \times 10^{22}$

Practice questions

3   a) 1.1345×10^4

b) 1×10^{-2}

c) 3.45×10^{-3}

4   a)   3 200 000

b)   845.6

c)   0.005 676 5

5   a) 1.132×10^{12}

b) 1.13×10^{-5}

c) 2.87×10^{15}

6   $1\,nm = 1 \times 10^{-9}\,m$

$0.070\,nm = 0.070 \times 10^{-9}\,m = 7.0 \times 10^{-11}\,m$

7   1.67×10^{-24}

8   amount (in moles) $= \dfrac{238}{119} = 2$

1 mole contains 6.02×10^{23} atoms

2 moles contain $2 \times 6.02 \times 10^{23}$ atoms $= 1.204 \times 10^{24}$

Fractions and percentages (pages 13–14)

Guided questions

1   Step 1 $M_r = 40 + (14 \times 2) + (16 \times 6) = 164$

Step 2 $14 \times 2 = 28$

Step 3 $\dfrac{\text{mass of nitrogen}}{M_r} = \dfrac{28}{164}$

Step 4 $\dfrac{28}{164} \times 100 = 17\%$

2   Step 1 $\dfrac{9}{24} = 0.375$

Step 2 $0.375 \times 6 = 2.25\,g$

Practice questions

3   $7.97 - 7.45 = 0.52\,g$

percentage increase $= \dfrac{0.52}{7.45} \times 100 = 6.98\%$

4   $\dfrac{30}{75} \times 100 = 40\%$

5   a) $\dfrac{2 \times 1}{74} \times 100 = 2.7\%$

b) $\dfrac{2 \times 39}{294} \times 100 = 26.6\%$

c) $\dfrac{2 \times 14}{132} \times 100 = 21.2\%$

6   $\dfrac{35}{100} \times 2.3 = 0.81\,g$

7   $\dfrac{4.1}{6.7} \times 100 = 61\%$

Ratios (pages 14–15)

Guided question

1 **Step 1**   P : O

   0.050 : 0.125

   Step 2 $\dfrac{0.050}{0.050} : \dfrac{0.125}{0.050}$

   1 : 2.5

   $1 \times 2 : 5 \times 2$

   2 : 5

   P_2O_5

Practice questions

2 a) $C_4H_5N_2O$

   b) $Na_2S_2O_3$

   c) CH_2O

   d) P_2O_5

3 a) Pb_3O_4

   b) Cl_2O_7

4   Pb : O

   $\dfrac{0.207}{207} : \dfrac{0.032}{16}$

   0.001 : 0.002

   1 : 2

   PbO_2

Balancing equations (page 16)

Guided question

1 **Step 1** $N_2 : H_2$

   1 : 3

   Step 2 There is three times as much H_2 as N_2, so divide H_2 moles by 3

   $\dfrac{0.4}{3} = 0.13$

Practice questions

2 a) $Cu(NO_3)_2 : O_2$

         2 : 1

         4 : 2 mol

   b) $Cu(NO_3)_2 : 4NO_2$

         2 : 4

         1 : 2

         $0.6 : 0.6 \times 2 = 0.12$ mol

3 a) CaO : C

         1 : 3

         $0.33 : 0.33 \times 3 = 0.99$ mol

   b) CaO : C

         1 : 3

         $3.2 : 3 \times 3.2 = 9.6$ mol

4 a) Pb : O_2

         3 : 2

         $1 : \dfrac{3}{2}$

         $0.66 : 0.66 \times \dfrac{2}{3} = 0.44$

   b) $2O_2 : Pb_3O_4$

         2 : 1

         $2.2 : \dfrac{2.2}{2} = 1.1$

   c) $3Pb : Pb_3O_4$

         3 : 1

         $0.33 : \dfrac{0.33}{3} = 0.11$

Estimating the results of simple calculations (page 17)

Guided question

1 **Step 2** Compound distance = 8 solvent distance = 20

   Step 3 Estimate the R_f value

   $\dfrac{8}{20} = \dfrac{4}{10} = 0.4$

   The R_f is approximately 0.4 so the compound is P

Practice questions

2 C

3 Anything in range +30 to +120

4 $100 \times 30 = 30\,000$

5 $10 \times 10 \times 10 = 1000\,mm^2$

Handling data

Significant figures (page 19)

Guided question

1 **Step 1** 3 478 906

   Step 2 34

   Step 3 The number after 7 is 8, which is greater than 5, so round up and make all the remaining digits zeros.

   Step 4 3 480 000

Practice questions

2 a) 4

   b) 4

   c) 5

   d) 4

   e) 2

3 a) 35 560

   b) 5.28

   c) 400

d) 442.5

e) 0.000045

4 $\% = \dfrac{2.53}{2.85} \times 100 = 88.8819 = 89\%$ (to 2 s.f.)

5 $95.0\,\text{cm}^3 = \dfrac{95}{1000} = 0.095\,\text{dm}^3$

$\text{moles of gas} = \dfrac{\text{volume of gas (dm}^3)}{24} = \dfrac{0.095}{24}$

$= 0.0039583$

$= 0.00396\,\text{mol}$

Reporting calculations to an appropriate number of significant figures (pages 20–21)

Guided question

1 **Step 1**

Measurement	Number of significant figures
$20.5\,\text{cm}^3$	3
$0.25\,\text{mol/dm}^3$	2
$1.2\,\text{mol/dm}^3$	3

Step 2 The least accurate measurement is to 2 significant figures and only 2 significant figures should be given in your final answer.

Step 3 $\text{moles NaOH} = \dfrac{\text{volume} \times \text{conc.}}{1000} = \dfrac{20.5 \times 0.25}{1000}$

$= 0.005125$

Step 4 moles HCl $= 0.005125$

Step 5 $\text{volume} = \dfrac{\text{moles} \times 1000}{\text{conc.}}$

$= \dfrac{0.005125 \times 1000}{1.2} = 4.2708$

Step 6 $4.3\,\text{cm}^3$ (to 2 s.f.)

Practice questions

2 The mass has 4 significant figures and the volume has 2 significant figures, hence the density should be given to 2 significant figures.

$\text{density} = \dfrac{\text{mass}}{\text{volume}} = \dfrac{40.52}{5.1} = 7.945098039$

$= 7.9\,\text{g/cm}^3$ (to 2 s.f)

3 $\text{amount (in moles)} = \dfrac{\text{mass (g)}}{A_r} = \dfrac{2.7}{40} = 0.0675$

$= 0.068\,\text{mol}$ (to 2 s.f.)

In the original data the mass was only given to 2 significant figures, hence the answer must be given to 2 significant figures.

4

Measurement	Number of significant figures
$26.50\,\text{cm}^3$	4
$0.200\,\text{mol/dm}^3$	3
$0.300\,\text{mol/dm}^3$	3

The least accurate measurement is to 3 significant figures and 3 significant figures should be given in your final answer.

$\text{moles NaOH} = \dfrac{\text{volume} \times \text{conc.}}{1000} = \dfrac{26.50 \times 0.200}{1000} = 0.0053$

ratio: 1 NaOH:1 HCl

$\therefore$ moles HCl $= 0.0053$

$\text{volume} = \dfrac{\text{moles} \times 1000}{\text{conc.}} = \dfrac{0.0053 \times 1000}{0.300} = 17.666$

$= 17.7\,\text{cm}^3$ (to 3 s.f)

Significant figures and standard form (page 22)

Guided question

1 **Step 3** 3.2090×10^3

Practice questions

2

Number	Number of significant figures	Number in standard form
0.0060	2	6.0×10^{-3}
50.08	4	5.008×10^1
3000.6g	5	3.0006×10^3
0.04070	4	4.070×10^{-2}

3 **a)** 5.0×10^{-2}

b) 1.2×10^{-2}

c) 1.230010×10^6

d) 1.4050×10^4

e) 3.003×10^1

Finding arithmetic means (pages 23–24)

Guided question

1 **Step 2** $27.05 + 27.15 + 27.15 = 81.35$

Step 3 $\dfrac{81.35}{3} = 27.12\,\text{cm}^3$

Practice questions

2 $\dfrac{35 + 36 + 37 + 37}{4} = 36.25 = 36\,°\text{C}$

3 River A: $\dfrac{14 + 13 + 11 + 9 + 8}{5} = 11$

River B: $\dfrac{8 + 9 + 10 + 11 + 9}{5} = 9.4$

River B is safest

4

	Titration 1	Titration 2	Titration 3	Titration 4
Initial burette reading/cm^3	0.00	14.00	0.00	15.30
Final burette reading/cm^3	13.00	26.50	12.45	28.00
Volume of HCl/cm^3	13.00	12.50	12.45	12.70

$\text{mean volume} = \dfrac{12.50 + 12.45}{2} = 12.50\,\text{cm}^3$

Calculating weighted means (page 25)

Guided question

1 **Step 1** $79 + 10 + 11 = 100$

Step 2

$$\text{relative atomic mass} = \frac{(79 \times 24) + (10 \times 25) + (11 \times 26)}{100}$$
$$= 24.32 = 24.3$$

Practice questions

2 $\frac{(69 \times 63) + (31 \times 65)}{100} = 63.62 = 63.6$

3 $\frac{(95.02 \times 32) + (0.76 \times 33) + (4.22 \times 34)}{100} = 32.09$

Bar charts and histograms (pages 26–27)

Guided question

1

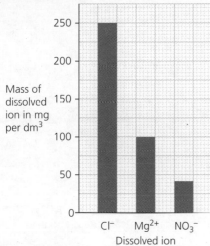

Practice question

2 **a)** aluminium

b) sodium

c) 2.3 mg/dm³

d) it is a gas and has a small mass per unit volume

e) it increases to aluminium, then decreases to phosophorus and increases to sulfur.

f) bar chart

Order of magnitude calculations (page 28)

Guided question

1 **Step 2** $\frac{0.118}{0.000\,001\,18} = 100\,000$ times bigger

Practice questions

2 $\frac{1 \times 10^{-4}}{1 \times 10^{-9}} = 1 \times 10^{5} = 100\,000$

3 $\frac{1.0 \times 10^{-6}}{1.6 \times 10^{-9}} = 0.625 \times 10^{3} = 625$ times bigger

Algebra

Understanding symbols (page 29)

Guided question
Step 2 $0.016 < 0.12$

Practice questions

1 **a)** False

b) True

c) True

d) False

e) False

Changing the subject of an equation (pages 30–31)

Guided questions

1 **Step 2** $3x + 6 - 6 = y - 3 - 6$
$$3x = y - 9$$

Step 3 $x = \dfrac{y - 9}{3}$

2 **Step 2**

$$\frac{(\text{volume} \times \text{conc.} \times \cancel{1000})}{\cancel{1000}} = \text{moles} \times 1000$$

$$\frac{\text{volume} \times \cancel{\text{conc.}}}{\cancel{\text{conc.}}} = \frac{\text{moles} \times 1000}{\text{conc.}}$$

$$\text{volume} = \frac{\text{moles} \times 1000}{\text{conc.}}$$

Practice questions

3 **a)** $x = \dfrac{y - 1}{2}$

b) $x = \dfrac{4 + y}{3}$

c) $x = 8 - y$

d) $x = \dfrac{y - c}{m}$

e) $x = \dfrac{3 - 2y}{4} = x$

f) $x = 1 - 2y$

4 **a)** $\text{mass} = \text{moles} \times M_r$

b) $\text{vol (dm}^3) = \text{moles} \times 24$

c) $\text{theoretical yield} = \dfrac{\text{actual yield} \times 100}{\text{percentage yield}}$

d) $\text{conc.} = \dfrac{\text{volume} \times 1000}{\text{moles}}$

e) $\text{time taken} = \dfrac{\text{quantity of reactant used}}{\text{mean rate of equation}}$

Substituting into algebraic equations and solving (pages 32–40)

Guided question

1   **Step 3** volume × conc. = moles × 1000

Step 4 conc. = $\dfrac{\text{moles} \times 1000}{\text{volume}}$

Step 5 conc. = $\dfrac{0.0034 \times 1000}{15.0}$ = 0.23 mol/dm³

Practice questions

2   $R_f = \dfrac{\text{distance moved by substance}}{\text{distance moved by solvent}}$

distance moved by solvent × R_f = distance moved by substance

10.2 × 0.80 = 8.2 cm

3   atom economy

$= \dfrac{\text{sum of relative formula mass of desired product from equation}}{\text{sum of relative masses of all reactants from equation}} \times 100$

$= \dfrac{160}{124 + 98} \times 100 = 72\%$

4   M_r KNO₃ = 39 + 14 + (3 × 16) = 101

moles × M_r = mass

0.45 × 101 = 45.45 g = 45 g (to 2 s.f.)

5   rate = $\dfrac{\text{quantity of reactant used}}{\text{time}} = \dfrac{0.20}{30} = 0.0067$ g/s

Graphs

Plotting graphs (pages 36–37)

Guided question

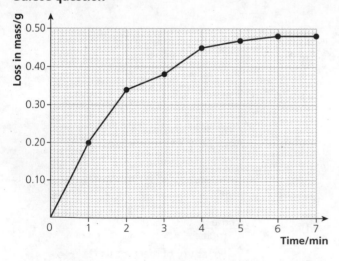

Practice questions

1

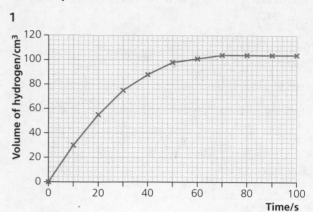

2

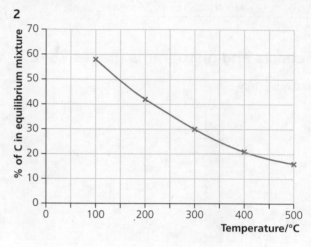

3   a)   Result at 4.5 minutes

b)   Mass of flask and contents /g; time /s

c)   A graph of mass of flask and contents against time for the reaction of calcium carbonate and acid

d)   Yes, as it fits most of the graph paper

A

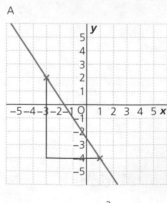

Slope = $-\dfrac{3}{2}$

$y = -\dfrac{3}{2}x - 2.5$

B

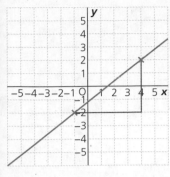

Slope = $\frac{4}{5}$

$y = \frac{4}{5}x - 1$

C

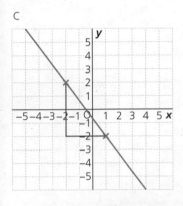

Slope = $-\frac{4}{3}$

$y = -\frac{4}{3}x - 0.5$

Determining the slope and intercept of a linear graph $y = mx + c$ (pages 39–40)

Guided question

1 a) **Step 3** Graph A $\Delta y = 2 - (-4) = 6$

Graph B $\Delta y = 2 - (-2) = 4$

Graph C $\Delta y = 2 - (-2) = 4$

Step 4 Graph A $\Delta x = -3 - 1 = -4$

Graph B $\Delta x = -1 - 4 = 5$

Graph C $\Delta x = -2 - 1 = 3$

Step 5 Graph A: gradient $= -1.5$

Graph B: gradient $= +0.8$

Graph C: gradient $= -1.3$

b) Graph A: intercept $= -2.5$;
equation: $y = -1.5 \times -2.5$

Graph B: intercept $= -1$; equation: $y = 0.8x - 1$

Graph C: intercept $= -0.5$; equation: $y = -1.3x - 0.5$

Practice questions

2 **A** positive   **B** negative   **C** positive   **D** zero
E negative   **F** negative   **G** positive

3 a) A 0.8   B −0.6   C 1.25   D −1.2   E 2.3   F 1.3

b) A −2   B −1   C −3   D −3   E −3   F 0

c) A $y = 0.8x - 2$ B $y = -0.6x - 1$

C $y = 1.25x - 3$ D $y = -1.2x - 3$

E $y = 2.3x - 3$ F $y = 1.3x$

Drawing and using the slope of a tangent to a curve (pages 42–43)

Guided question

1 **Step 3** gradient $(m) = \dfrac{\text{change in } y\text{-axis}}{\text{change in } x\text{-axis}} = \dfrac{\Delta y}{\Delta x} = \dfrac{1}{2}$

Practice questions

2

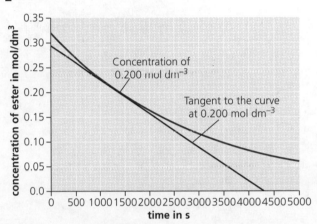

At a concentration of 0.200 mol/dm³ the gradient of the tangent is

$\dfrac{0.28}{4300} = 6.5 \times 10^{-5}$

3 a)

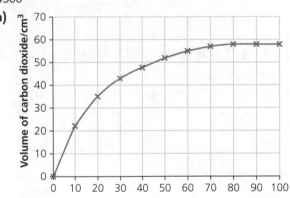

b) gradient of tangent $= \dfrac{30}{28} = 1.1\,\text{cm}^s/\text{s} = \text{rate}$

c) gradient of tangent $= \dfrac{10}{34} = 0.3\,\text{cm}^s/\text{s} = \text{rate}$

Geometry and trigonometry

Representing 2D and 3D (pages 45–46)

Guided question

1
H—N—H
 |
 H

Practice questions

2
 H
 |
H—C—H
 |
 H

3
 H
 |
H—C—O—H
 |
 H

4
 O
 / \
H   H

5 **a**
 H   H
 | |
H—C—C—H
 | |
 H   H

b
H   H
| |
C=C—C—H
| | |
H   H   H

c
 H   H   H
 | | |
H—C—C—C=C
 | | | |
 H   H   H   H

Area, surface area and volume (page 47)

Guided question

1 **Step 2** surface area : volume

24 : 8

3 : 1

Practice questions

2 A is 1 : 1.5

B is 1 : 1.9

C is 1 : 3.5

so the biggest ratio is C

3 surface area is $2 \times 2 \times 6 = 24\,\text{cm}^2$

volume $= 2 \times 2 \times 2 = 8\,\text{cm}^3$

surface area : volume

24 : 8

3 : 1

4 A: surface area $= 5 \times 5 \times 6 = 150\,\text{cm}^2$

volume $= 5 \times 5 \times 5 = 125\,\text{cm}^3$

surface area : volume

150 : 125

30 : 25

6 : 5

B: surface area $= 3 \times 3 \times 6 = 54\,\text{cm}^2$

volume $= 3 \times 3 \times 3 = 27\,\text{cm}^3$

surface area : volume

54 : 27 (divide by 3)

18 : 9

2 : 1

cube B has the bigger ratio

5 surface area $= 2 \times 2 \times 6 = 24\,\text{cm}^2$

volume $= 2 \times 2 \times 2 = 8\,\text{cm}^3$

surface area : volume

24 : 8

3 : 1

surface area $= 20 \times 20 \times 6 = 2400\,\text{cm}^2$

volume $= 20 \times 20 \times 20 = 8000\,\text{cm}^3$

Surface area : volume

2400 : 8000 (divide by 3)

300 : 1000

3 : 10

0.3 : 1

The cube with small sides has a surface area to volume ratio, which is ten times bigger.

›› 2 Literacy

Extended responses: 'Describe' (pages 52–53)

Expert commentary

1 This is a model answer that would get full marks.

Measure out 25 cm³ of sulfuric acid in a measuring cylinder and place in a beaker. Warm the acid and add some spatulas of zinc carbonate to the acid and you will see bubbles and the zinc carbonate disappears. Keep adding the zinc carbonate until there is some solid on the bottom of the beaker. Now filter the mixture to separate the excess solid from the solution and place the solution in an evaporating basin. Heat the solution in the evaporating basin using a Bunsen burner until half the water has evaporated. Allow to cool and crystallise filter if necessary and dry the crystals between sheets of filter paper or in a desiccator.

Peer assessment

2 This would be awarded a Level 2 and given 3 marks.

The student obtained Level 2 because a description is given that demonstrates a reasonable knowledge and understanding of some elements of the titration and preparation of crystals. There are some inaccuracies and some steps are missing. The description is mostly coherent and logical. Three marks were awarded due to the large number of omissions.

The inaccuracies are:

- Universal indicator should not be used in a titration – methyl orange (which changes from yellow to red) or phenolphthalein (which changes to pink to colourless) should be used instead. The student did, however, give the correct indicator colour change for universal indicator.

- The technique of making a titration accurate by rinsing apparatus and swirling was not included.

- The volume of acid used in the titration should be recorded and the titration repeated without using indicator, but adding the same volume of acid, this means the crystals will be pure and will not contain the indicator.

- The solution should be cooled to crystallise before drying.

Improve the answer

3 A model answer that would be awarded the full marks is:

I would place some zinc chloride in an evaporating basin. I would put two electrodes in the zinc chloride and attach one to the positive (the anode) of a power pack or battery and the other to the negative (the cathode). I would place the evaporating basin on a gauze and heat gently with a Bunsen burner. When the zinc chloride is beginning to melt, I would make sure the electrodes do not touch, and then switch on the power pack. At the positive anode a grey liquid is observed and at the cathode a green-yellow gas. A fume cupboard should be used.

Extended responses: 'Explain' (pages 54–55)

Expert commentary

1 A model answer that would be awarded the full marks is:

Copper is a metal and has lots of delocalised electrons which can move and carry charge and so it conducts electricity. Copper chloride is an ionic compound. It cannot conduct when it is solid because the ions are held tightly in place but when it is dissolved in solution then the ions can move and carry the charge. Chlorine is a molecule and it does not have charge and so it cannot conduct electricity.

Peer assessment

2 This would be awarded a Level 3 and given 5 marks.

This is a Level 3 answer because a detailed and coherent explanation is given that demonstrates a good knowledge and understanding of all of the physical properties. It is logically presented.

There are a few minor errors:

- Graphite has 'a strong bond between metals and non-metals', this should state that there is a strong bond between carbon atoms.

- 'Graphite has a giant covalent bond'. Giant covalent is a structural term, and instead the student should state that graphite has covalent bonds between atoms.

- A better answer would have mentioned that the delocalised electrons can move.

Improve the answer

3 This is an improved answer that would get all 6 marks.

In sodium chloride there are sodium ions and chloride ions that have strong ionic bonds between them. Therefore it takes a large amount of energy to break these strong bonds in sodium chloride. Chlorine is a small molecule and there are weak intermolecular forces between the molecules. These forces are broken when the substance melts. These weak intermolecular forces do not need much energy to break and so chlorine therefore has a low melting point.

Extended responses: 'Plan/Design' (pages 56–57)

Expert commentary

1 This is a model answer that would get full marks.

First, measure out 25 cm^3 using a measuring cylinder of 0.1 mol/dm^3 hydrochloric acid and place it in a conical flask. Quickly start a timer and add 0.05 g of magnesium ribbon. Stop the timer and record the time when there are no more bubbles and the reaction is finished. Repeat the experiment using the same mass of magnesium and a different concentration of hydrochloric acid, for example 0.2 mol/dm^3. Repeat again using three other different concentrations of hydrochloric acid and the same mass of magnesium. The temperature would be kept the same by immersing the flask in a thermostatically controlled water bath.

Peer assessment

2 This would be awarded a Level 2 and given 3 marks.

This is a Level 2 answer because a plan is given that demonstrates a reasonable knowledge and understanding but lacks some practical detail and does not fully identify the compound.

The inaccuracies include:

- The method of carrying out a flame test and correct colours of the flame tests were not included.

- The sulfate test is correctly described, but the test for the chloride and bromide ion is incorrect. Silver nitrate solution should be added after the nitric acid, and the colours of the precipitates are the wrong way around.

Improve the answer

3 This is an improved answer that would get all 6 marks.

First place some potassium iodide solution in a test tube and add some aqueous chlorine. If there is a colour change to a yellow/brown solution then iodine has been produced because chlorine is more reactive and will displace iodine from the solution.

In a second test tube, place some potassium iodide solution and add some aqueous bromine. If the solution turns to yellow/brown then iodine has been produced because bromine is more reactive and will displace iodine from the solution.

In a third test tube, place some potassium bromide solution and add some aqueous chlorine. If the solution turns orange/red brown then bromine has been produced because chlorine is more reactive than bromine and will displace it from the solution.

Extended responses: 'Evaluate' (pages 58–59)

Expert commentary

1 This is a model answer that would get full marks.

The raw material to make hydrogen is water and there is an abundant supply, for example in seas and lakes. Diesel comes from crude oil, which is in limited supply and is running out. To save crude oil, a non-renewable resource, it is better to use hydrogen. Hydrogen is expensive to produce from water as electricity is needed and the generation of electricity may produce carbon dioxide that contributes to the greenhouse effect unless renewable power is used.

When hydrogen burns it produces water only and so it does not cause any air pollution, but diesel burns to produce carbon dioxide that can cause the greenhouse effect. The greenhouse effect causes global warming and ice caps to melt. Incomplete combustion of diesel may produce carbon monoxide, which is toxic, and also carbon, which can cause smog that causes respiratory problems. In conclusion, hydrogen is better to use because it is in good supply and does not cause pollution, but it is a flammable gas, which is expensive to store safely.

Peer assessment

2 This would be awarded a Level 2 and given 3 marks.

This is Level 2 because an attempt has been made to describe some conditions that comes to a conclusion for temperature. However, the logic may be inconsistent at times, particularly in terms of pressure, but it does build towards a coherent argument for temperature conditions.

The inaccuracies are:

- The student stated that increasing pressure did not have much effect, however, increasing pressure increases the yield.

- The student did not comment on the pressure to be used.

Improve the answer

3 This is an improved answer that would get all 6 marks.

The plastic carton is made from crude oil, which is a finite resource. The cardboard carton is made from wood which is a renewable resource. The temperature to produce the plastic carton is twice as high as that to produce cardboard and so more energy is needed. If this energy is generated using non-renewable fuels then this will harm the environment. There is a greater mass of carbon dioxide produced per kilogram and this can lead to global warming and melting of polar ice caps. The cardboard carton may be more biodegradable than the plastic carton.

Extended responses: 'Use' (pages 60–62)

Expert commentary

1 This is a model answer that would get full marks.

The table shows that the Group 2 elements react more vigorously as the group descends. Group 1 elements have the same trend. Group 1 elements include lithium, sodium and potassium and they all react with cold water. However, the table shows that one element in Group 2, beryllium, does not react with water but the other elements in Group 2 do. This is different for Group 1 as all the elements of Group 1 react with cold water.

The table shows that for Group 2 the reaction changes from no reaction to a slow, moderate, rapid and then very rapid reaction. This is the same for Group 1. The table shows that the Group 2 metals that react hydrogen gas always produce a hydroxide. This is the same for Group 1 metals.

Peer assessment

2 This would be awarded a Level 3 and given 5 marks.

This is a Level 3 answer because a detailed and coherent explanation is given that demonstrates a good knowledge and understanding and it refers to energy changes and temperature changes for all three substances, correctly deducing that a negative sign means heat is given out.

The inaccuracies include:

- The student writes that the temperature got colder. It would be more accurate to write that the temperature decreases.

- The last sentence about time is irrelevant.

Improve the answer

3 This is an improved answer that would get all 6 marks.

From the table I can deduce that the halogens react more slowly and less vigorously with iron as the group descends. This means that the reactivity of the halogens decreases down the group. This is because halogen atoms have seven electrons in the outer shell and they need to react to gain one electron to obtain a full outer shell. The further down the group, the further the electron is from the nucleus as the atoms get bigger. This means that the electron gained is less strongly attracted to the nucleus and so harder to gain. The harder the electron is to gain, the less reactive the halogen.

» 3 Working scientifically

The development of scientific thinking (pages 69–70)

1 Ethanol is flammable so do not heat it directly, instead use a hot water bath.

2 a) Carrying out an experiment; drawing conclusions/reporting results.

b)

Risk	Control measure
Potassium nitrate powder is an irritant	Wear gloves/wash hands immediately if some falls on skin

3 a) Cut the potassium into small pieces; use a large volume of water; use tweezers to hold the potassium; use a safety screen and wear goggles – any **three** from list.

b) The speed of reaction of the metal depends on the reactivity of the metal.

4 a) It prevents tooth decay.

b) They are against mass medication; and there is a lack of freedom of choice.

5 C

6

Risk	Control measure
Ethanol is flammable	Use a water bath to heat
Conc. sulfuric acid is corrosive	Wear gloves and goggles

7 The results should be peer assessed. This means they should be evaluated by other scientists working in the same field.

Experimental skills and strategies (pages 72–73)

1 a) independent: volume of hydrochloric acid

dependent: temperature

control: mass and surface area of magnesium; concentration of acid

b) independent: mass of copper carbonate

dependent: time taken to react and disappear

control: surface area of copper carbonate; volume and concentration of acid; temperature

c) independent: type of acid used

dependent: temperature

control: volume of sodium hydroxide

2 a) speed of dissolving depends on temperature

b) blue solution formed

c) beaker/test tube; stirring rod; stopwatch/balance

3 The student should add missing part of glassware attached to bung to the diagram – copper(II) sulfate solution in a (conical) flask/boiling tube attached to glassware with the bung

Pure water in test tube/flask/beaker at end of delivery tube, which must not be sealed

Heat source to heat container holding copper(II) sulfate solution

4 a) +/– 0.01

b) units of g for the second column

5 a) magnesium + copper sulfate → magnesium sulfate + copper

b) The student has stated what has happened in the experiment but has not given observations. The correct observations are red brown solid forms or the blue solution changes in colour to colourless.

6 a) any **two** from:
- limewater turns cloudy
- solid/calcium carbonate disappears
- bubbles/fizzing
- mass decreases

explanation: the bubbles/mass decrease/limewater cloudy is because carbon dioxide gas is released

b) 0.1 g

7

Time /s	10	20	30	40	50
Volume 1 /cm³	30	49	59	63	63
Volume 2 /cm³	32	51	59	63	65
Average volume /cm³	31	50	59	63	64

Analysis and evaluation (page 76)

Questions

1 a) Student A = 21.3 ± 0.2%, Student B = 22.5 ± 0.1%

b) Student A was accurate, Student B was not accurate

c) both students had repeatable results

d) due to random errors

e) Student B had a systematic error; results were consistently about 1.3% too high

2 a) the result at 4.5 minutes

b) 100.3 g

c) as time increases the mass of the flask and contents decreases. The mass decreases more rapidly from 103.0

to 99.4g between 0 and 4 minutes, then it gradually decreases until at 7 minutes, when it is constant at 99.0g

d) repeat more times and calculate the mean

e) use a different balance

3 a) systematic error

b) random error

c) systematic error

≫ Exam-style questions

Paper 1 (pages 104–108)

1 a) 2 marks for plotting points correctly, 1 mark for a smooth line. [3]

b) The temperature increases so heat was given out. [1]

c) Burette [1]

d) Use a lid on the cup with a hole, to allow the burette in [1], this prevents heat loss [1].

Or

Use a pipette to measure the acid [1]. It is accurate to one decimal place [1].

e) $NaOH + HCl$ [1] $\rightarrow NaCl + H_2O$ [1]

f) $\frac{25.0 \times 0.1}{1000}$ [1] $= 0.0025\,mol$ [1]

g) It is completely ionised in aqueous solution. [1]

h) Add an indicator [1] and record the temperature at the colour change [1].

i) $M_r\ NaOH = 40.$
concentration $= \frac{40.0}{40} \times \frac{1000}{250}$ [1] $= 4\,mol/dm^3$ [1]

2 a) bubbles (due to the carbon dioxide produced) [1]

b) Award 1 mark for each of the indicative content points covered below, up to a maximum of 6 marks.
Indicative content
- measure out a volume of dilute sulfuric acid and place in conical flask/beaker
- add excess magnesium carbonate to dilute sulfuric acid
- filter to remove excess magnesium carbonate
- heat the filtrate to half volume evaporate some water/heat to the point of crystallisation
- leave to cool so crystals form
- dry between pieces of filter paper or in a low temperature oven

c) Examples may include: Wear safety glasses [1]; tie long hair back [1].

d) $M_r\ MgCO_3 = 84, M_r\ MgSO_4 = 120$ [1]
$\frac{2.1}{84} = 0.025\,mol\ MgCO_3$ [1]
$0.025 \times 120 = 3.0\,g\ MgSO_4$ [1]

e) $\frac{1.8}{3.0} \times 100$ [1] $= 60\%$ [1]

f) Some may be lost in filtering or during transfer between apparatus. [1]

3 a) phenolphthalein (colourless to pink)/methyl orange (yellow to red) [1]/litmus (blue to red)[1]

b) colour change would not be visible in red wine. [1]

c) i $18.90 + 19.00/2$ [1] $= 18.95\,cm^3$ [1]

ii $18.95 \times \frac{0.100}{1000}$ [1] $= 0.001895\,mol\ NaOH$ [1]

2 mol NaOH : 1 mol tartaric acid

$\frac{0.001895}{2} = 0.0009475\,mol\ tartaric$ [1]

$0.000975 \times \frac{1000}{25.0}$ [1] $= 0.379 = 0.038\,mol/dm^3$ [1]

iii 150 [1]

4 Award 1 mark for each of the indicative content points covered below, up to a maximum of 6 marks.

Indicative content
- thermometer
- measuring cylinder/pipette
- thermometer
- spatula
- plastic cup (with lid)
- weigh the same mass of each metal in each same state of division, e.g. powder
- measure a volume of sulfuric acid into a plastic cup
- measure and record the temperature of the sulfuric acid
- add metal P into the plastic cup
- stir and record the highest temperature
- repeat for each metal at least 3 times to calculate a mean
- calculate the mean temperature change, the greatest temperature change gives the most reactive metal

5 a) A: Gas syringe [1]; B: Conical flask [1]

b) $Mg + 2HCl \rightarrow H_2 + MgCl_2$ [2]

c) Advantage: convenient/quick to use [1]

Disadvantage: inaccurate/only accurate to $1\,cm^3$ [1]

d) Award marks as follows: sensible scales, using at least half the grid for the points [1], all points correct [1], best-fit line [1]

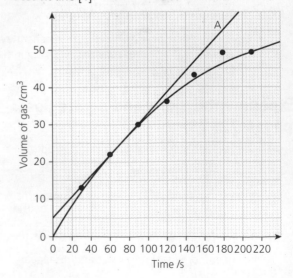

e) 80 (read from graph) [1]

f) Volume of gas $= 13\,cm^3$ [1]

$$\text{mean rate} = \frac{13}{30} = 0.433\,cm^3/s \quad [1]$$

$$= 0.4\,[1]\,cm^3/s \text{ (to 1 s.f.) } [1]$$

g) Correct tangent drawn, see part d [1]

$$\frac{36}{120} = 0.3 \text{ [1] (depending on graph)}$$

$$= 3 \times 10^{-1}\,[1]\,cm^3/s\,[1]$$

h) Line is steeper and to the left of original [1]; line finishes at same overall volume of gas collected [1]

i) There would be a faster reaction at $40\,°C$ [1]; particles have more energy, so more successful collisions with the required activation energy per second. [1]

6 a) total mass $= 37.5\,g$ [1]

$$\% \text{Sn} = \frac{15.0}{37.5} \times 100 \text{ [1]} = 40\% \text{ [1]}$$

b) C

7 a) So that it does not run on the paper [1]

b) three [1]

c) distance moved by $X = 3.3$ cm [1]; distance moved by solvent $= 4.4$ cm [1]; R_f value $= \frac{3.3}{4.4} = 0.75$ [1]

8 Award 1 mark for each of the indicative content points covered below, up to a maximum of 6.

Indicative content
- phytomining uses plants to absorb copper compounds from the ground
- the plants are harvested and then burned
- an ash is produced that contains the copper compounds
- acid is added to form a leachate solution
- bioleaching uses bacteria to produce leachate solutions that contain copper compounds
- copper can be obtained from solutions of the copper compounds by displacement using scrap iron, $Fe + CuSO_4 \rightarrow Cu + FeSO_4$

9 a) $6CO_2 + 6H_2O \rightarrow C_6H_{12}O_6 + 6O_2$ for $C_6H_{12}O_6$ balancing [1]

b) carbon dioxide dissolves in seawater and used by sea life to form calcium carbonate/shells [1], which becomes sediment and forms sedimentary rock [1].

c) i it has increased [1]

ii $\frac{(390 - 370)}{370} \times 100$ [1] $= \frac{20}{370} \times 100 = 5.41\%$ [1]

iii increased burning of fossil fuels [1], deforestation [1]

10 Award 1 mark for each of the points below, up to a maximum of 4 marks.

- In Earth's atmosphere there is much less carbon dioxide – 0.04%. [1]
- In Earth's atmosphere there is much more nitrogen – 78%. [1]
- In Earth's atmosphere there is much more oxygen – 21%. [1]
- In Earth's atmosphere there is no methane. [1]
- Both have noble gases in small amounts. [1]

Key terms

Accurate: Something that is close to the true value and repeatable.

Anhydrous: A substance that does not contain water.

Arithmetic mean: The sum of a set of values divided by the number of values in the set – it is sometimes called the average.

Avogadro's number: The number of atoms, molecules or ions in one mole of a given substance.

Bar charts: Charts showing discrete data in which the height of the unconnected bars represents the frequency.

Categoric variable: A variable with a value that is a label, e.g. the name of an acid.

Concordant results: Concordant means exactly the same, so concordant results are the exact same readings.

Continuous data: Data which can take any value within a range, such as the mass of a beaker.

Continuous variable: A variable that has a numerical value.

Control (or controlled) variable: The variable that is neither dependent or independent, and that must be kept constant during an experiment.

Dependent variable: The variable that is measured whenever there is a change in the independent variable.

Error: The difference between an observed value and what is true in nature. Error causes results that are inaccurate or misleading.

Empirical formula: The simplest whole number ratio of atoms present.

Ethics: The consideration of the moral right or wrong of an action.

Evaluate: This means to weigh up the good points and the bad points.

Histograms: Charts showing continuous data in which the area of the bar represents the frequency.

Hydrated: A substance that contains water.

Hypothesis: A proposal intended to explain certain facts or observations.

Independent variable: The variable that is deliberately changed in an experiment.

Key word: A word that helps you communicate ideas in science clearly.

Meniscus: The curve that is seen at the top of a liquid close to the surface of the container.

Model: A representation of a thing or process in a way that aids understanding.

Order of magnitude: If we write a number in standard form, the nearest power of 10 is its order of magnitude.

Peer review: The evaluation of scientific work by others working in the same field.

Precise: A set of results that are close to the mean value where there is little spread around a mean value.

Qualitative: Descriptions of how something appears, rather than with figures or numbers.

Quantitative: Measurements such as mass, temperature and volume involve a numerical value. For these quantitative measurements, it is essential that units are included because stating that the mass of a solid is 0.4 says very little about the actual mass of the solid – it could be 0.4 g or 0.4 kg.

Random error: An error that causes a measurement to differ from the true value by different amounts each time.

Relative formula mass M_r: The sum of the relative atomic masses (A_r) of all the atoms shown in the formula.

R_f: R_f value in chromatography is the distance travelled by a given component divided by the distance travelled by the solvent.

Repeatable: A set of results that can be obtained by the same person repeating an experiment.

Reproducible: A result that can be repeated by another person or using a different technique.

Resolution: The smallest change a piece of apparatus can measure.

Risk assessment: A judgement on how likely it is that someone might come to harm if a planned action is carried out and how these risks could be reduced.

Significant figures: Approximations to a number, determined by a set of mathematical rules.

Systematic error: An error that causes a measurement to differ from the true value by the similar amounts each time.

Theory: A hypothesis that has been proved by experiment or observation to be correct.

Uncertainty: The range of measurements within which the true value can be expected to be.

Valid results: Results or data obtained from an appropriately designed experiment.

Variable: This is a characteristic or chemical quantity.

Command words

Calculate: Use numbers given in the question to work out the answer.

Choose: Select from a range of alternatives.

Compare: Describe the similarities and/or differences between things, not just write about one.

Complete: Answers should be written in the space provided, for example on a diagram, in spaces in a sentence, or in a table.

Define: Specify the meaning of something.

Describe: Recall some facts, events or processes in an accurate way.

Design: Set out how something will be done.

Determine: Use given data or information to obtain and answer.

Draw: To produce or add to a diagram.

Estimate: Assign an approximate value.

Evaluate: Use the information supplied, as well as knowledge and understanding, to consider evidence for and against.

Explain: Make something clear or state the reasons for something happening.

Give: Only a short answer is required, not an explanation or a description.

Identify: Name or otherwise characterise.

Justify: Use evidence from the information supplied to support an answer.

Label: Provide appropriate names on a diagram.

Measure: Find an item of data for a given quantity.

Name: Only a short answer is required, not an explanation or a description. Often it can be answered with a single word, phrase or sentence.

Plan: Write a method.

Plot: Mark on a graph using data given.

Predict: Give a plausible outcome.

Show: Provide structured evidence to reach a conclusion.

Sketch: Draw approximately.

Suggest: Apply knowledge and understanding to a new situation.

Use: The answer must be based on the information given in the question. Unless the information given in the question is used, no marks can be given. In some cases, you may have to use your own knowledge and understanding.

Write: Only a short answer is required, not an explanation or a description.